MANAGING THE COLUMBIA RIVER

INSTREAM FLOWS, WATER WITHDRAWALS, AND SALMON SURVIVAL

Committee on Water Resources Management, Instream Flows, and Salmon Survival in the Columbia River Basin

Water Science and Technology Board

Board on Environmental Studies and Toxicology

Division on Earth and Life Studies

NATIONAL RESEARCH COUNCIL
OF THE NATIONAL ACADEMIES

THE NATIONAL ACADEMIES PRESS
Washington, D.C.
www.nap.edu

THE NATIONAL ACADEMIES PRESS 500 Fifth Street, N.W.
Washington, DC 20001

NOTICE: The project that is the subject of this report was approved by the Governing Board of the National Research Council, whose members are drawn from the councils of the National Academy of Sciences, the National Academy of Engineering, and the Institute of Medicine. The members of the committee responsible for the report were chosen for their special competences and with regard for appropriate balance.

Support for this project was provided by the Washington State Department of Ecology under Contract No. SLOC C0300043. Any opinions, findings, conclusions, or recommendations expressed in this publication are those of the author(s) and do not necessarily reflect the views of the organizations or agencies that provided support for the project.

Additional copies of this report are available from the National Academies Press, 500 Fifth Street, N.W., Lockbox 285, Washington, DC 20055; (800) 624-6242 or (202) 334-3313 (in the Washington metropolitan area); Internet, http://www.nap.edu.

International Standard Book Number 0-309-09155-1 (Book)
International Standard Book Number 0-309-53037-7 (PDF)

Copyright 2004 by the National Academy of Sciences. All rights reserved.

Printed in the United States of America.

THE NATIONAL ACADEMIES
Advisers to the Nation on Science, Engineering, and Medicine

The **National Academy of Sciences** is a private, nonprofit, self-perpetuating society of distinguished scholars engaged in scientific and engineering research, dedicated to the furtherance of science and technology and to their use for the general welfare. Upon the authority of the charter granted to it by the Congress in 1863, the Academy has a mandate that requires it to advise the federal government on scientific and technical matters. Dr. Bruce M. Alberts is president of the National Academy of Sciences.

The **National Academy of Engineering** was established in 1964, under the charter of the National Academy of Sciences, as a parallel organization of outstanding engineers. It is autonomous in its administration and in the selection of its members, sharing with the National Academy of Sciences the responsibility for advising the federal government. The National Academy of Engineering also sponsors engineering programs aimed at meeting national needs, encourages education and research, and recognizes the superior achievement of engineers. Dr. William A. Wulf is president of the National Academy of Engineering.

The **Institute of Medicine** was established in 1970 by the National Academy of Sciences to secure the services of eminent members of appropriate professions in the examination of policy matters pertaining to the health of the public. The Institute acts under the responsibility given to the National Academy of Sciences by its congressional charter to be an adviser to the federal government and, upon its own initiative, to identify issues of medical care, research, and education. Dr. Harvey V. Fineberg is president of the Institute of Medicine.

The **National Research Council** was organized by the National Academy of Sciences in 1916 to associate the broad community of science and technology with the Academy's purposes of furthering knowledge and advising the federal government. Functioning in accordance with general policies determined by the Academy, the Council has become the principal operating agency of both the National Academy of Sciences and the National Academy of Engineering in providing services to the government, the public, and the scientific and engineering communities. The Council is administered jointly by both Academies and the Institute of Medicine. Dr. Bruce M. Alberts and Dr. William A. Wulf are chair and vice-chair, respectively, of the National Research Council.

www.national-academies.org

COMMITTEE ON WATER RESOURCES MANAGEMENT, INSTREAM FLOWS, AND SALMON SURVIVAL IN THE COLUMBIA RIVER BASIN[*]

ERNEST T. SMERDON, *Chair*, University of Arizona (Emeritus), Tucson
RICHARD M. ADAMS, Oregon State University, Corvallis
DONALD W. CHAPMAN, Consultant, McCall, Idaho
DARRELL G. FONTANE, Colorado State University, Fort Collins
ALBERT E. GIORGI, BioAnalysts, Inc., Redmond, Washington
HELEN M. INGRAM, University of California, Irvine
W. CARTER JOHNSON, South Dakota State University, Brookings
JOHN J. MAGNUSON, University of Wisconsin, Madison
STUART W. McKENZIE, U.S. Geological Survey (Retired), Gresham, Oregon
DIANE M. McKNIGHT, University of Colorado, Boulder
TAMMY J. NEWCOMB, Michigan Department of Natural Resources, Lansing
KENNETH K. TANJI, University of California (Emeritus), Davis
JOHN E. THORSON, Attorney-at-Law, Oakland, California

National Research Council Staff

JEFFREY W. JACOBS, Study Director, Water Science and Technology Board
DAVID POLICANSKY, Associate Director, Board on Environmental Studies and Toxicology
ELLEN A. DE GUZMAN, Research Associate, Water Science and Technology Board

---

[*] This committee was organized, overseen, and supported by the Water Science and Technology Board (lead) and the Board on Environmental Studies and Toxicology. Biographical sketches of committee members are contained in Appendix E and rosters of the two parent boards are contained in Appendix F.

Preface

The Columbia River and its basin have long comprised one of the great natural resources of the United States. For thousands of years, salmon from the river provided an important food resource for Native Americans, as the river dependably produced vast amounts of salmon to be eaten fresh or dried, which ensured adequate levels of dietary protein. As the United States was developing and expanding westward in the early nineteenth century, President Thomas Jefferson promoted exploration of the recently-acquired lands of the Louisiana Purchase. This led to the 1804-1806 expedition of Lewis and Clark to explore and chronicle the American West and to pursue Jefferson's goal of finding the fabled water route to the Pacific. The expedition's voyage to the Pacific Ocean took them upstream on one of the nation's great rivers, the Missouri, then downstream on another of the nation's great rivers, the Columbia. Since then, the vast and diverse resources of the Columbia River basin have been utilized and have contributed to the region's economic and population growth, which gained momentum many years ago and continues today.

Efforts at harvesting the resources of the Columbia in the late nineteenth and early twentieth centuries including irrigated agriculture on the basin's arid, but fertile, lands, as well as a commercial fishing industry. Low-cost hydroelectric power attended and aided agricultural development, which included cities beyond the Columbia's drainage basin. The urban corridor from north of Seattle to south of Portland and beyond continues to grow, and this human population growth puts ever-increasing social, political, and economic pressures on the resources of the Columbia River. It also increases tensions among the various enterprises that desire a greater portion of the river's largesse.

In the meantime, the salmon populations of the Columbia

River have been steadily declining since the first dam was built on the river. In fact, several species of salmon are now listed as endangered under the federal Endangered Species Act. By law, efforts must be made to protect these species from further degradation and to start the process of recovery. The dilemma is how to protect the salmon and the Columbia River's natural resources and still enable those resources to be used to further enrich the region's economy. A key issue in this study was the pending applications for additional water rights permits from the mainstem Columbia River in the State of Washington, applications which have been on hold for some time. Our committee's charge was to consider the implications for potential additional withdrawals for Columbia River salmon and to comment on the body of scientific knowledge related to this issue and its implications. The committee was not charged to review all ecological issues (of which there are many) across the basin which affect salmon but rather to conduct a more focused investigation regarding conditions in a stretch of the mainstem Columbia River in the State of Washington. Nor was the committee charged with recommending policy decisions but rather was requested to review the scientific information available by decision makers and to comment on it.

To address these issues, the Washington State Department of Ecology requested the National Research Council (NRC) to conduct a study addressing specific issues given in the Statement of Task provided in the body of this report. The committee avoided the temptation to go beyond the tasks assigned—although each member, while not encumbered by biases or personal gain from any direction the study might take, nonetheless had personal views, some strongly held. All members had experience that related to one or more aspects of the issues at hand. The committee strove to ensure that the many viewpoints expressed by committee members were heard before coming to a consensus on what should be included in the report. The resulting report represents the collective view of the committee. In some cases it may differ from what individual members might have written. The composition of the committee was such that most disciplines related to the issues contained in the charge to the committee were represented by experienced and knowledgeable people. I thank the committee members, all of whom volunteered many hours of personal time without financial compensation. Their reward is the sense of satisfaction in objectively addressing a problem of importance to all citizens of the Columbia River basin, the larger Pacific Northwest, and the nation.

The committee devoted a great deal of time at its meetings listening to interest group representatives, scientists, as well as private citizens, to learn more about the broad range of interests and concerns regarding the Columbia River and its salmon. Still, one group central to the committee's task did not speak—the various species of salmon, whose populations have been in general decline since the introduction of an industrial-based economy. But several people we visited with spoke on behalf of the salmon and on related environmental issues.

Our committee is grateful to the Washington State Department of Ecology for its insight regarding the need for an objective, independent look at issues related to survival of the various salmon species and how water management decisions in the Columbia River basin might affect the fate of salmonids. We thank Tom Fitzsimmons, Gerry O'Keefe, and their colleagues at the Department of Ecology who provided support and assistance before and during our study. We also thank all members of a "Resources Group," which consisted of several experienced expert scientists from the region. The Department of Ecology invited these experts to provide input to this study. The committee found the presentations from these experts, which were provided in open public meetings in early 2003, extremely useful and informative.

The committee held four meetings in 2003, the first three in the State of Washington and the last at the NRC in Washington, D.C. The process involved presentations at the first two meetings from the Department of Ecology and its staff, the Resources Group, and others with specific interests or expertise. All information-gathering meetings were open and publicly announced. The committee sought to hear from as many groups and individuals as was possible within the time constraints, and all speakers and guests were invited to provide written extensions of their comments at the meeting or subsequent to it. All presentations and written comments were carefully considered by the committee. The committee thanks all individuals who provided oral and/or written information, as that information was very helpful. The committee's final two meetings were held in closed session without guest speakers or other visitors so that the committee could focus on its deliberations related to its Statement of Task and its draft report.

The committee and particularly I as committee chair thanks the NRC staff members for their dedication and diligent work in making this report highly professional. I particularly thank Jeffrey Jacobs, senior staff officer with the Water Science and Technology Board, who laboriously pored over lengthy and often too verbose input to put together a concise and coherent report. Jeff and the committee were ably assisted by Ellen de Guzman, research associate with the board, who handled administrative details for the meetings and ably assisted in all phases of report preparation. Finally, David Policansky, associate director of Board on Environmental Studies and Toxicology, provided input and guidance, attending all meetings and contributing to the committee's deliberations. This report is the work of the committee in terms of scientific input, but the final professional product is due to the efforts of the NRC staff.

This report was reviewed in draft form by individuals chosen for their diversity of perspectives and technical expertise in accordance with procedures approved by the NRC's Report Review Committee. The purpose of this independent review was to provide candid and critical comments to assist the institution in making its published report as sound as possible and to ensure that the report meets institutional standards for objectivity, evidence, and responsiveness to the study charge. The review comments and draft manuscript remain confidential to protect the integrity of the deliberative process. We thank the following reviewers for their helpful suggestions, all of which were considered and many of which were wholly or partly added to the final report: Ellis Cowling, North Carolina State University (emeritus); William Kirby, U.S. Geological Survey; Ronald Lacewell, Texas A&M University; Pamela Matson, Stanford University;

Willis McConnaha, Mobrand Biometrics; Kathleen Miller, National Center for Atmospheric Research; William Pearcy, Oregon State University; Brian Richter, The Nature Conservancy; Will Stelle, Preston Gates; John Williams, NOAA Fisheries; Robert Wissmar, University of Washington; and Ellen Wohl, Colorado State University. Although these reviewers provided many constructive comments and suggestions, they were not asked to endorse the conclusions or the recommendations, nor did they see the final draft of the report before its release. The review of this report was overseen by Robert Beschta, Oregon State University, appointed by the NRC's Division on Earth and Life Studies, and Stephen Berry of the University of Chicago, appointed by the NRC's Report Review Committee. They were responsible for ensuring that an independent examination of the report was carefully carried out in accordance with NRC institutional procedures and that all review comments were considered. Responsibility for the final content of this report rests entirely with the authoring committee and the NRC.

The Department of Ecology faces great challenges in addressing the complex issues of managing Columbia River resources in the State of Washington. It must work with the other basin states, one Canadian province, several Native American tribes, and other interested entities. It will face many political pressures. But we are sure of its sincerity in finding a balance so that no interest is ignored, even if compromise is required by all. We wish the department the best of luck as it faces these challenges, and we hope this report is useful in formulating future Columbia River basin decisions and policies.

Ernest T. Smerdon,
Chair

Contents

EXECUTIVE SUMMARY 1

1
INTRODUCTION 15
Columbia River Salmon 15
Study Background, Process, and Organization 23

2
DEVELOPMENT AND CHANGES IN THE COLUMBIA RIVER BASIN 27
Settlement and Development of the Columbia Basin 28
Federal Columbia River Hydropower System 36
Summary 39

3
HYDROLOGY AND WATER MANAGEMENT 42
Columbia River Flows 46
Water Withdrawals 52
Return Flows and Water Quality 60
Water Temperature 63
Climate Variability and Change 65
Summary 69

4
ENVIRONMENTAL INFLUENCES ON SALMON 71
Columbia River Salmon 72
Status of Salmon and Steelhead Stocks 76
Research, Modelling, and Alternative Hypotheses 83
Water Temperature and Flow Management 96
Summary 104

5
WATER LAWS AND INSTITUTIONS 107
Introduction 107
International Obligations 108
Interstate Compacts 111
Interstate Apportionment 114
Native American Water and Fisheries Rights 116
Federal Rights and Obligations 129
State Laws and Institutions 133
Summary 144

6
BETTER MANAGEMENT OF EXISTING WATER
SUPPLIES 146
The Economic Value of Water 146
Water Markets and Water Banks 158
Summary 172

7
WATER RESOURCES MANAGEMENT, RISKS, AND
UNCERTAINTIES 175
Risk and Water Management 175
Columbia River Management Decisions 182
The Management Scenarios 188
Summary 194

8
EPILOGUE 199

REFERENCES 203
APPENDIXES
A Columbia River Initiative Draft Management Scenarios, July 7, 2003 223
B Resources Group 229
C Calculations on Annual Discharges of Water From the Columbia Basin Project 230
D Climate Change and Hydrologic Impacts 235
E Committee Biographical Information 238
F National Research Council Board Membership and Staff 244

Executive Summary

BACKGROUND

For thousands of years, North America's Columbia River salmon runs were the most abundant on Earth. The salmon evolved in a setting of many long- and short-term environmental changes and disruptions. With the introduction of an industrial-based economy to the region in the late nineteenth century, the scale and rate of environmental variability in the basin changed. The creation of impoundments on the Columbia River and its tributaries, dam operations, commercial fishing, logging, diversions for irrigated agriculture, and human population growth have altered the Columbia's presettlement flow regime and have reduced the quality of salmon habitat across the river basin. There have been attendant declines—including some extinctions—in the populations of all resident salmon species. Many of these salmon are currently listed as threatened and endangered pursuant to the federal Endangered Species Act. Annual salmon and steelhead returns to the Columbia River estuary are estimated to have been as high as 16 million fish per year during the late 1800s. The returns have dwindled over time, dropping to near 1 million fish per year in the 1990s. These numbers rebounded in the late 1990s and early 2000s, largely because that time frame coincided with a period of favorable ocean conditions for salmon. The majority of returns today consist of hatchery-reared fish.

The Columbia River makes up part of a large (basin size of roughly 250,000 square miles) ecological system with many features that vary naturally on several different timescales. In addition to natural ecological variability, salmon are affected by human-induced changes such as water diversions and water control structures. Furthermore, Columbia River salmon spend most of their lives in the highly dynamic Pacific Ocean. The combination of these and other factors presents a setting of extraordinary variability and uncertainty for Columbia River salmon. The life cycles of Columbia River salmon (there are several different species and subspecies) have been intensively studied. In fact, Columbia River salmon are among the world's most carefully studied fish species, and this research has yielded an excellent understanding of salmon physiology and migratory behavior.

The Washington State Department of Ecology issues water use permits for the portion of the Columbia River that flows through the state. Water withdrawal permit decisions must be balanced with the state's obligation to protect and enhance the quality of the natural environment, including salmon habitat. The department considers scientific knowledge of salmon and environmental variables in making permitting decisions. That body of knowledge, as extensive and thorough as it may be, is imperfect and contains competing theories, models, and perspectives.

This is the context in which the department requested that the National Research Council (NRC) provide advice regarding salmon and water management decisions. In response to this request, the NRC reviewed and evaluated existing scientific data and analyses related to fish species listed under the Endangered Species Act in the Columbia River basin and reviewed and evaluated environmental parameters critical to the survival and recovery of listed fish species. The cumulative effects and the risks to survival of listed fish species of potential future water withdrawals of between approximately 250,000 acre-feet and 1.3 million acre-feet per year were also evaluated. There are currently many pending water withdrawal permit applications along the Columbia River in the State of Washington. The total volume of water represented by these applications falls within this 250,000 to 1.3 million acre-feet per year range. In addition, the effects of proposed management criteria, specific diversion quantities, and specific features of potential water management alternatives provided by the state were also considered. To conduct the study, the NRC appointed the ad hoc Committee on Water Resources Management, Instream Flows, and Salmon Survival in the Columbia River. This report's Preface contains additional information about the study process, and Chapter 1 includes verbatim the committee's Statement of Task.

SALMON AND ENVIRONMENTAL PARAMETERS

There are competing scientific hypotheses and models regarding the effects of environmental forces on Columbia River salmon. River velocity and water temperature are of particular interest to fisheries scientists, water managers, and interest groups, as these factors influence the migratory behavior of salmonids. Several computer models have been used to simulate the effects of river flows (especially water velocity) and temperature on the migratory speed and survival of smolt (young salmon ready to migrate from fresh water to the sea). These models ascribe different levels of importance to river discharge and temperature and their effects on migratory conditions for juvenile salmonids. Selecting the "best" model of salmon-environmental relationships was neither part of this study nor was it critical to its completion. Several scientists presented analyses and models in open public meetings for consideration in this study. These presentations were used as background information for considering the degree to which proposed future water extractions may pose increased risks to the survival of endangered fish species. This information, along with the body of scientific evaluations of Columbia River salmon and their habitat, portrays a complex system of interacting environmental variables that influence the rates of salmon smolt survival on their downstream journey through the Columbia River hydrosystem. **Within the body of scientific literature reviewed as part of this study, the relative importance of various environmental variables on smolt survival is not clearly established. When river flows become critically low or water temperatures excessively high, however, pronounced changes in salmon migratory behavior and lower survival rates are expected.**

COLUMBIA RIVER FLOWS AND WITHDRAWALS

Changes to the Annual Hydrograph

The annual flow patterns of the Columbia River underwent a substantial transformation during the twentieth century. At the beginning of the century, the river's flows exhibited great seasonality, with roughly 75 percent of the annual flows occurring during summer months (April-September) and roughly 25 percent of annual flows occurring during winter months (October-March). The river's long-term average discharge is roughly 139 million acre-feet per year. The pattern of annual flows changed in response to the construction of numerous mainstem and tributary impoundments and the subsequent operations of this water control system. The system is known as the Federal Columbia River Power System (FCRPS), and the principal original purposes underlying its construction were to provide hydroelectricity, irrigation, and flood control benefits. Construction of some of the system's large mainstem projects, such as Grand Coulee and Bonneville dams, began in the 1930s. The post-World War II period saw a burst in project authorization and construction of additional large projects. Other projects were built in connection with the Canada-U.S. Columbia River Treaty signed in 1961. The hydrological implications of the system's construction were tremendous. As the system's water control projects came on line, annual flows of the Columbia became and less and less seasonal, as the differences between summer and winter flows were reduced in order to provide reliable year-round hydropower generation and distribution. In the late 1970s, the Columbia's annual flows had been modified such that they were divided roughly evenly between summer and winter, as compared to the 75:25 ratio that had existed at the beginning of the twentieth century. In addition to this "flattening" of the annual Columbia River hydrograph, other key impacts of the construction and operations of the hydropower system were a decrease in water velocities, a change in the size and orientation of the river's plume (a physical zone in the Pacific Ocean that extends from the Columbia's mouth into marine waters), and major changes to limnology and nutritional pathways in the river's estuary and food web. All these changes have likely had significant effects on the early ocean survival of juvenile fish leaving the Columbia River. Passage of such legislation as the National Environmental Policy Act (1969) and the Endangered Species Act (1973) resulted in changes in operational patterns and priorities. "Flow targets" have been established by federal and state agencies in an effort to sustain and recover salmon habitat and populations that had declined over time. The FCRPS today is operated primarily to provide benefits of flood control, hydropower, and instream flows.

This study's focus was on the implications of potential additional water withdrawals (which would be primarily for irrigated agriculture) from the mainstem Columbia River for salmon survival. The study charge did not call for an examination of the hydrological impacts of consumptive withdrawals in comparison with other actions, such as the creation of impoundments, dam operations, or changes in land cover. Knowledge of these historical changes to Columbia River hydrology, however, provided important context for the consideration of the specific issues within this study's Statement of Task.

Prospective Additional Water Withdrawals

Of special interest in this study was consideration of the effects and risks to salmonid survival of a specific range of possible additional water withdrawals, ranging from 250,000 acre-feet per year to 1.3 million acre-feet per year. The latter figure represents roughly 28 percent of the total volume of water permits that have been issued to the present by the State of Washington for surface water withdrawals from the Columbia River and groundwater withdrawals from the zone within 1 mile of the river. The effects of these proposed withdrawals and their attendant risks for the survival of a specific species will vary considerably depending on river flow levels. Despite construction and operations of the hydropower system, the river still exhibits considerable flow variations on daily, seasonal, and annual timescales. Under current conditions, less than 1 percent of total annual withdrawals are made during January. By contrast, during July—the month of highest withdrawals—about 18 percent of annual withdrawals from the Columbia River in the State of Washington are made. The seasonality of water withdrawals is of utmost importance when considering how the river's water withdrawals affect salmon survival rates.

Many calculations and speculations could be made with regard to the range of prospective additional withdrawals considered in this study. Assuming that the monthly pattern of withdrawals from the mainstem Columbia River continues essentially unchanged and that the maximum amount of prospective withdrawals in the range considered in this study (maximum of 1.3 million acre-feet per year) is diverted, additional withdrawals of roughly 2,600 acre-feet in January and roughly 234,000 acre-feet in July would result. The effects of these prospective additional January withdrawals would result in additional withdrawals of less than 1 percent of mean January Columbia River flow. The effects of these prospective additional withdrawals in July, when river flows are lower, would increase July withdrawals from their current value of roughly 6.8 percent of mean flows to roughly 8.6 percent of mean flows. Under *minimum* July flow conditions, the effects would be even greater: the upper end of the proposed range of diversions would increase current July withdrawals from roughly 16.6 percent to roughly 21 percent of Columbia River *minimum* flows. Water temperature is also a concern to salmon survival. Columbia River water temperatures have been increasing for decades, and those temperatures are at their highest during summer months (when demand for extractions is also at or near its peak). Water quality is also an issue, as return flows from irrigated agriculture and urban activities are of degraded quality and could affect fish already stressed from higher water temperatures and longer travel times.

The scale of the Columbia River basin and current limits of scientific understanding of salmon and their habitat inhibit reliable, precisely quantified predictions of how additional water withdrawals will affect risks to salmon survival. Nevertheless, further reductions in river flows during low-flow periods will increase those risks, especially since most of those withdrawals would occur during a critical period for those salmon species that are migrating through the mainstem river. There are differences in the migration patterns and timing of the Columbia River's listed salmon species and subspecies. Accordingly, only those salmon populations that migrate (downstream or upstream) through the river corridor during critical low-flow periods or years will be exposed to the greater risks entailed by additional withdrawals and reductions in discharge. Examples of these populations include subyearling ocean-type Chinook from the Snake and Columbia rivers, adult Snake and Columbia River summer Chinook, adult Snake and Columbia River steelhead, and adult sockeye salmon.

Columbia River salmon today are at a critical point. The basin's salmon populations have been in steady decline over the past century, and scientific evidence demonstrates that environmental and biological thresholds important to salmon—such as water temperature—are being reached or in some cases exceeded. Salmon are more likely to be imperiled during late summer on the Columbia River, as they experience pronounced changes in migratory behavior and survival rates when river flow becomes critically low or water temperature becomes too high. Further decreases in flows or increases in water temperature are likely to reduce survival rates. Trends such as human population growth in the region and prospective regional climate warming further increase risks regarding salmon survival.

Decisions regarding the issue of additional water withdrawal permits are matters of public policy, but if additional permits are issued, they should include specific conditions that allow withdrawals to be discontinued during critical periods. Allowing for additional withdrawals during the critical periods of high demand, low flows, and comparatively high water temperatures identified in this report would increase risks of survivability to listed salmon stocks and would reduce management flexibility during these periods.

WATER MANAGEMENT INSTITUTIONS

A Joint Forum for Considering Water Withdrawal Applications

The Columbia River basin is a single hydrological unit extending over seven U.S. states, many Indian reservations, and one Canadian province. Water permitting decisions are made by basin states with few obligations or attempts to make those decisions in a spatially coordinated manner across the entire basin. This fragmented basis for making water rights permitting decisions represents a barrier to better decision making in this realm. It also inhibits consideration of the cumulative effects of additional small individual withdrawals. The effects of any one newly authorized individual water withdrawal from the Columbia River on flows and temperature are likely to be minimal. The effects of additional small diversions accumulate, however, and will eventually have serious consequences for salmon, especially when interacting with such variables as climate, ocean conditions, and human population growth. The current "case-by-case" approach for evaluating the effect of water permits on salmon can be likened to a beaver felling a tree—the effect of any single wood chip removed by the beaver on the health of the tree is slight and indeterminable. Critical thresholds, however, are crossed as the tree is girdled, reducing growth and causing mortality of major branches, or eventually removing enough wood to fell the tree. Every bite has only a small effect in itself, but each one contributes to the tree's eventual felling. Columbia River salmon are being subjected to a similar process. In isolation, small additional water withdrawals each have an imperceptible effect on survival rates of salmon; but the cumulative effects of many small additional individual water withdrawals throughout the river's basin collectively could push salmon across life-threatening thresholds, particularly in critical periods of high demand and low flows.

Decreases in river flows have been caused by one very large diversion along the river—the long-approved large diversions for the Columbia Basin Project clearly dominate historical diversions—along with a large number of small individual actions. A process in which water rights permitting applications throughout the basin are considered apart from this phenomenon of cumulative effects has contributed to declining salmon populations and may be contributing to political tensions. Decisions regarding prospective additional diversions should be considered with an understanding of existing and potential future diversions across the entire basin and should be subjected to professional and public scrutiny, and consideration of risk factors and systemwide equities. The lack of such a basinwide framework also tends to discourage efforts at conservation and better management, since such measures employed in one state or other entity will have limited usefulness if other states and entities do not enact similar measures.

The State of Washington and other basin jurisdictions should convene a joint forum for documenting and discussing the environmental and other consequences of proposed water diversions that exceed a specified threshold. This forum could be convened within the existing Northwest Power and Conservation Council, which includes broad representation of political entities from across the basin. The council has accomplished good things, and discussions of water permit applications could be integrated into its resource management responsibilities. Limitations of convening this forum within the council include possible administrative and legal complications of extending the council's functions. Convening the forum within a new simple framework could offer the advantage of greater flexibility and a clearer focus of responsibilities and obligations.

Better Management of Existing Water Supplies

Water management approaches such as water conservation and associated transfers, conjunctive use of groundwater, water markets, water banks, and environmental water accounts have the potential to support regional economic growth without requiring additional Columbia River water diversions. They are also likely to require investments in physical infrastructure and in human resources. These approaches can help transfer water between willing buyers and willing sellers and can be useful in helping shift water in response to changing economic conditions and priorities as well as during periods of shortage. Physically, they may entail transfers of water in conveyance facilities or the storage of water in a reservoir or groundwater reserve to be used later during a period of high demand. In some cases they may require the construction of conveyance and storage facilities. These approaches can be important in promoting a prosperous Columbia River basin economy that meets human needs while sustaining viable salmon populations and a healthy river ecosystem. Water supplies procured through these means could augment both water deliveries and instream flows. To be effective, such systems must consider and devise safeguards for preventing undue harm to third parties. **The State of Washington and other Columbia River basin entities should continue to explore prospects for water transfers and other market-based programs as alternatives to additional withdrawals.**

MAKING COLUMBIA RIVER MANAGEMENT DECISIONS

Washington State Department of Ecology Water Management Scenarios

The water management scenarios proposed by the Washington State Department of Ecology and that were considered in this study contained many assumptions and actions related to water withdrawal quantities, management actions, and water use fees (key features of the scenarios, and comments that resulted from this study, are listed below; Appendix A lists these scenarios in their entirety). Some of the scenarios promote adaptive management concepts, which is appropriate and encouraging. Several possible management actions did not contain enough specificity to enable detailed evaluation. A pervasive aspect of the scenarios is the lack of comprehensive, basin-wide consideration of water uses and needs as a context for evaluating withdrawal permit applications.

Key features of the scenarios, along with commentary and evaluation, are listed below.

- *Conversion of interruptible to uninterruptible water rights (Scenarios 1-4).*

The needs of some users (especially growers of perennial crops) for uninterruptible withdrawals are understandable. The downside of such a system, however, is that uninterruptible status makes adaptive responses in periods of stress more difficult. Uninterruptible water rights are pre-1980 state law water rights that have priority over mainstem instream flow rights that were established in 1980. These rights stand in contrast to interruptible water rights, which may be curtailed under certain low-flow conditions to protect mainstem instream flows.

The conversion of water rights to uninterruptible status will decrease the flexibility of the system during critical periods of low flows and comparatively high water temperatures. Conversions to uninterruptible rights, during these critical periods, are not recommended.

- *Criteria for state-of-the-art efficiency (Scenarios 1-4).*

The criteria for assessing the state-of-the-art (water use) efficiency measures are not described. In addition, organizational responsibility for making that evaluation is not specified.

- *Reevaluation at 10 and 20 years (Scenarios 1-3).*

The idea of reevaluating the scenarios periodically is excellent and is consistent with adaptive management principles. For this reevaluation to be meaningful, decisions should be able to be adjusted if evaluation calls for such. No evidence of any such reversibility was provided. In some cases, more frequent reevaluations may be in order.

- *Monitoring and metering (Scenarios 1-3).*

Monitoring for compliance with standards and water metering are excellent ideas and could be accomplished as a part of this report's recommended basinwide joint forum for discussing

Columbia River basin water permit applications.

- *Charges for water rights (Scenarios 2-4).*

Charges for water rights appear to be arbitrarily chosen and out of proportion to the probable costs of mitigation and the value of water. For example, Scenario 2 specifies a charge of $10 per acre-foot per year to be used (among other things) to acquire mitigation water in low-water years. Even in high-water years, the economic value of out-of-stream water is greater than $10 per acre-foot per year, and this value increases in low-water years. This scenario seemingly poses selling water rights for $10 per acre-foot per year, when water may later have to be purchased for several times that amount.

- *Water markets.*

Proposals within the scenarios to establish water markets and water banks are appealing, as they offer potential improvements over existing water allocation systems. However, restricting markets to the Columbia River mainstem, and only to the State of Washington, is narrowly construed. For example, the Department of Ecology already allows for 600,000 acre-feet per year to be used by Oregon, but no allowance is made for uses in Canada, Idaho, or Montana, or by tribal groups. Efforts toward developing water markets should be complemented with efforts to evaluate third-party effects and to design proposals for compensating users indirectly harmed by water rights transfers.

- *Structural storage measures.*

Structural measures imply that tributaries are to be used for additional storage, but ecological habitat and conditions in tributaries are important for many reasons, including their relationship to Columbia River salmon survival. Tributaries should be considered for protection and for mitigation as well.

- *Scenario 5.*

This scenario was labeled a "no action" scenario, yet it prescribes new actions in that it allows for additional water withdrawal permits. The notion of consulting with fishery managers is good; however, no mention is made of criteria for the evaluation, how the results of the evaluation might be enforced, who decides how much mitigation is needed, and what—if any—limits on new permits might be enacted.

- *Mitigation.*

Mitigation measures are suggested in most of the management scenarios. Although the idea of "mitigating" impacts is attractive, the reality of most mitigation measures is that they are not well coordinated; that is, a management agency may attempt to offset harmful impacts of water withdrawals in one part of a river system with mitigation measures (e.g., ecosystem restoration) elsewhere. The ultimate outcomes of such varying actions, however, are difficult to accurately predict, measure, and compare (if indeed they are ever measured and meaningfully compared, which they often are not), thus making it difficult to determine if "mitigation" was actually achieved.

Science and Decision Making

The management of Columbia River salmon is an exceedingly complex public policy issue. The creation of comprehensive management strategies that enhance viable salmon populations, that calm disputes, and that meet human and economic demands will likely require a flexible and collaborative decision-making approach that involves scientists, managers, and decision makers. Science has contributed greatly to the collective knowledge of Columbia River salmon, but "better" or "more" scientific information will not necessarily lead to the resolution of disputes or to better management decisions. **Sound, comprehensive salmon management strategies will depend not only on science but also on a willingness by elected and duly appointed leaders and managers to take actions in the face of uncertainties**. It will also depend on scientists and managers working in a process in which managers and elected officials help frame scientific investigations and inquiry. The scientific knowledge of Columbia River salmon is as extensive as for any fish species in the world. Improvements in salmon habitat and return rates will require a willingness to employ existing scientific knowledge—despite its imperfections—to address some of the factors that scientific research suggests have led to their declines. A process in which scientists monitor the outcomes of management actions and provide feedback to stakeholders and decision makers (who then adjust management actions accordingly—generally referred to as "adaptive management") will be instrumental in helping understand how additional scientific research can best support management decisions.

1

Introduction

The Columbia River and its tributaries constitute one of North America's great river systems (Figure 1-1). The Columbia River Basin extends over an area of 258,000 square miles (Leopold, 1994), covering portions of seven U.S. states and one Canadian province. The river stretches 1,214 miles from its source in the mountains of the Canadian province of British Columbia to the Pacific Ocean. One of the Columbia's main tributaries is the Snake River, which drains most of the basin's southeastern reaches and enters the Columbia near the Tri-Cities (Kennewick, Pasco, and Richland) region of central Washington. Other important tributary streams are the Clearwater, Deschutes, Kootenai, Pend Oreille, Salmon, and Willamette rivers.

COLUMBIA RIVER SALMON

The Columbia River is well known for its rich variety of salmon species and populations. Columbia River salmon once existed in great abundance and for thousands of years served as the foundation of the diets of the region's Native American tribes. Lewis and Clark described the abundance of Columbia River salmon during their expedition to the region in 1805 to 1806:

> Captain Clark . . . halted at two large mat-houses. Here, as at the three houses below, the inhabitants were occupied in splitting and drying salmon. The multitudes of this fish are almost inconceivable. The water is so clear that they can readily be seen at the depth of 15 or 20 feet; but at this season they float in such quantities down the stream, and are drifted ashore, that the Indians have only to collect, split, and dry them on the scaffolds. (Coues, 1893, p. 641)

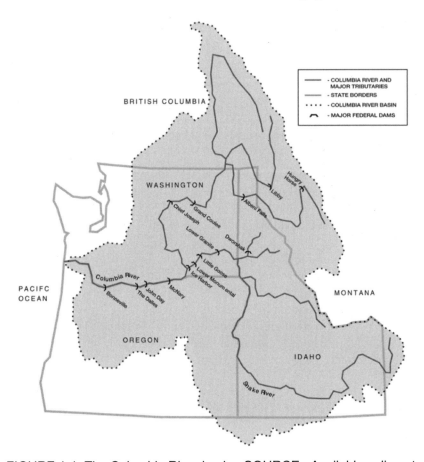

FIGURE 1-1 The Columbia River basin. SOURCE: Available online at http://www.bpa.gov/power/pg/fcrps_brochure_17x11.pdf, last accessed June 18, 2004..

The Pacific Northwest and its salmon populations and habitat have undergone many changes in the 200 years following Lewis and Clark's transcontinental adventure. The region has experienced substantial human population growth, and attendant land use changes have altered vegetation and hydrological patterns. Hydropower dams on the Columbia mainstem and hundreds of storage, diversion, and smaller-scale hydropower dams on its tributaries have altered the volume, velocity, and seasonality of river flows. The cumulative effects of these and other changes have contributed to a long-term decline in the number of

Introduction

View of Bonneville Dam, with its spillway and two powerhouses. Photo courtesy of Jeffrey Jacobs.

adult salmon returning to the river to spawn. Historic annual runs of salmon and steelhead, believed to have been at times as great as 16,000,000 fish (NPCC, 1986), declined to about 1,000,000 by the 1990s (*http://www.nwppc.org/library/pocket guide/pocketguide.pdf*, last accessed November 20, 2003), and increased in the late 1990s and early 2000s.

Six species of anadromous salmonids inhabit the Columbia basin: (1) Chinook salmon (*Oncorhynchus tshawytscha*); (2) coho, or silver, salmon (*O. kisutch*); (3) chum salmon (*O. keta*); (4) sockeye or red salmon (*O. nerka*); (5) pink or humpback salmon (*O. gorbuscha*); and (6) steelhead (*O. mykiss*). Chinook, coho, sockeye, and steelhead that migrate through the middle and upper reaches (above Bonneville Dam) of the Columbia and Snake rivers are all listed as federally endangered species. Columbia River salmon and steelhead stocks that are "threatened" or "endangered" (Table 1-1) under the Endangered Species Act include:

TABLE 1-1 Federally Threatened and Endangered Columbia River Salmonid Species

Endangered Species
Steelhead
  Upper Columbia River Steelhead
Chinook Salmon
  Upper Columbia River Spring Run Chinook
Sockeye
  Snake River Sockeye

Threatened Species
Steelhead
  Snake River Basin Steelhead
  Lower Columbia River Steelhead
  Middle Columbia River Steelhead
  Upper Willamette River Steelhead
Chinook Salmon
  Snake River Spring/Summer Chinook
  Snake River Fall Chinook
  Upper Willamette Chinook
  Lower Columbia River Chinook
Coho Salmon
  Lower Columbia River/Southwest Washington Coho (candidate for listing)
Chum Salmon
  Columbia River Chum Salmon

SOURCE: Data from NOAA Fisheries, National Marine Fisheries Service. Available online at *http://www.nmfs.noaa.gov/prot_res/species/ESA_species.html*, last accessed February 6, 2004.

1. Snake River fall Chinook salmon, threatened (Snake River upstream from Lyons Ferry Hatchery to Hells Canyon Dam, including lower reaches of the Clearwater, Imnaha, Grande Ronde, Salmon, and Tucannon rivers).

2. Snake River spring/summer Chinook salmon, threatened (wild/natural spawners in several subbasins of the Snake and Salmon rivers, including tributaries of the lower Snake River, Grande Ronde and Imnaha rivers, South Fork Salmon River, Middle Fork Salmon River, and the Upper Salmon River).

3. Mid-Columbia River steelhead, threatened (tributaries in the Columbia Plateau region, including Rock Creek, Fifteenmile Creek, and the White Salmon, Klickitat, Yakima Deschutes, John Day, Umatilla, and Walla Walla rivers).

4. Snake River sockeye salmon, endangered (Redfish Lake

in the upper Salmon River basin).

5. Upper Columbia River spring Chinook salmon, endangered (tributaries upstream from Rock Island Dam, including the Wenatchee, Entiat, and Methow rivers).

6. Upper Columbia River steelhead, endangered (tributaries upstream from Rock Island Dam, including minor tributaries to the Columbia River and the Wenatchee, Entiat, and Methow rivers).

Other Native and Exotic Fish Species

The Columbia River, like many western U.S. rivers, has far fewer native fish species than similar-sized rivers in the central and eastern United States. Before the construction of dams, the native fauna was dominated by salmonids (salmon and trout), cyprinids (minnows), and cottids (sculpins), most of which are still present but often in reduced numbers. In addition to the salmonids mentioned above, the basin also supports populations of bull trout (*Salvelinus confluentus*) and coastal cutthroat trout (*O. clarki*). Bull trout are federally listed as threatened and are found throughout the Columbia River basin. They typically reside in upper-tributary streams, reservoirs, and lakes and are found occasionally in the mainstem Columbia River. The salmonid complex also included whitefishes and ciscoes. The largest cyprinid in the Columbia River is the northern pikeminnow (formerly known as the northern squawfish). The white sturgeon is usually anadromous (spending part of its life in fresh water and part in saltwater), but landlocked populations also inhabit the Columbia River basin (Lee et al., 1980). By the late twentieth century, the white sturgeon population had declined to a point at which they were no longer considered commercially viable in the lower Columbia River (Craig and Hacker, 1940). White sturgeon are today found in small numbers in distinct landlocked populations. Several species of lamprey also exist in the Columbia River. Counts of lampreys reveal greatly diminished populations (CRITFC, 1996), and there have been efforts to classify the species as threatened or endangered.

Exotic or nonnative fish species have been widely introduced into the western United States and the Columbia River basin is no exception. The striped bass and the American shad, native to the eastern United States, are nonnative anadromous species that inhabit the Columbia River. Other nonnative freshwater fish in the Columbia River system are largemouth bass, smallmouth bass, sunfish, crappie, walleye, carp, catfish, bullhead, brown trout, brook trout, and lake trout. Many of these species have thrived in the altered conditions of the Columbia River system; however, some may have been more productive in an undammed river. Many of these species prey on salmon eggs and fry, and some—especially larger individuals—eat salmon juveniles as well (Zimmerman, 1999). Among nonnative species, walleye have particularly been implicated in connection with reduced salmon population productivity. Smallmouth bass and channel catfish also prey on salmon, and these predators are more abundant in the upper Columbia and Snake rivers than in the lower Columbia (ibid).

Commercial Fishing

The earliest commercial activity in the Columbia basin may have been fishing, as Native American tribes caught and traded salmon products. The introduction of commercial fishing, processing, and distribution practices to the region in the late nineteenth century resulted in a burst of economic activity and the generation of a great deal of wealth. The intensity of commercial fishing eventually led to declines in salmonid fish populations in the early- to mid-twentieth century. According to one estimate, total salmon harvest peaked at an average of approximately 34,000,000 pounds per year between 1880 and 1930, then declined to 24,000,000 pounds per year in the 1940s, dropped to 11,000,000 pounds per year in the 1950s, and was recorded at 1,200,000 pounds per year in the early 1980s (Huppert and Fluharty, 1995). Today, little commercial fishing is allowed or even possible given the small stock sizes in the basin (ibid.). Between 1981 and 1993, total average annual landed value for commercial fisheries in the basin was about $6.8 million (1993 dollars). Roughly half of the commercial value is generated in connection with fishing allowed under the American Indian Treaty, which is based on the 1974 ruling by a federal judge regarding the balance between tribal and nontribal fishing (Chapter 5 discusses the *Boldt* decision and other Native American water issues).

From 2000 to 2003 large runs of Chinook increased the commercial value of harvested salmon. However, the listing of several salmon and steelhead stocks in the Columbia River as threatened or endangered during the 1980s and 1990s triggered provisions of the Endangered Species Act, which in turn has limited opportunities for increased salmon harvests from more abundant (nonlisted) stocks. Chinook and coho salmon dominate the commercial fishing catch (93 percent), with white sturgeon also being important. Employment in fisheries is today relatively small, and in 1995 the salmon component was negligible because of low stock abundance. Depending on the assumption of annual income levels, the number of jobs currently dependent on the basin fisheries is estimated to be between 200 and 400 (Huppert and Fluharty, 1995). Although this is a small percentage of regional employment, commercial fishing has great economic importance in some local areas and communities. The value of commercial fishing has been declining, but fishing remains a major component of the region's recreation and tourism sector. The value of recreational fishing (mostly steelhead) is estimated at $7.7 million annually (1993 dollars). Recreational fishing is enjoyed throughout the basin, particularly downstream of Bonneville Dam. Important localized fisheries occur upstream from Bonneville Dam for fall chinook and for hatchery spring/summer chinook and steelhead. Catch-and-release fisheries for steelhead in some tributaries are also locally important. Furthermore, nonconsumptive fishery-based recreation, such as viewing salmon spawning in rivers and streams, and viewing fish at dams, hatcheries, and fish ladders, generates an estimated $80 million a year in expenditures (Huppert and Fluharty, 1995). In some areas, entire communities, resorts, businesses, and individuals depend heavily on services related to recreational fishing.

Salmon Management and Science

Identifying appropriate operational responses to facilitate the recovery of salmon populations is a complex scientific and policy task. Dozens of federal, state, and local organizations are responsible for managing the river, its extensive system of dams, and land uses across the watershed. For decades, many scientists and science organizations have investigated the varied aspects of salmon issues. These issues are complicated by the fact that the salmon are anadromous, spending part of their lives in freshwater and part in saltwater. Moreover, the salmon's habitat extends beyond the Columbia River basin. They pass downstream through the Columbia River estuary, spend one to five years (depending on the species) in marine residence, and then return to their natal streams to spawn. Clear understanding of how additional water withdrawals are likely to affect salmon species and their habitat is thus precluded by many factors. In addition to these factors, smolt (young salmon two or three years old that have acquired a silvery color) survival rates are affected by factors beyond streamflow seasonality and discharge, including water temperature, water chemistry, and changes in both land use and the estuarine environment. Existing scientific research and predictive models provide only partial information on these complex relationships and how they might change in the future. Moreover, there are competing models and paradigms, and not all scientists agree on the fundamental relationships among such parameters as flow, temperature, predation, and salmon survival rates.

Several important scientific issues in Columbia River management revolve around the relationships between resident juvenile salmon, smolts, survival rates, and instream flows. These issues are especially important on the middle reach of the river in central Washington, where the Washington State Department of Ecology is responsible for the water rights permitting process. Washington State water law is based on the western U.S. doctrine of prior appropriation, in which water rights are required to make withdrawals. The permitting agency must consider several factors in deciding whether to issue water rights to new, or "junior," appropriators, including possible impacts of additional withdrawals on federally endangered salmon species. Applicants for new water rights would like to receive permits in order to support economic activities and growth; however, additional withdrawals may negatively affect survival rates of salmon smolts.

Introduction

STUDY BACKGROUND, PROCESS, AND ORGANIZATION

The ambiguities and tensions surrounding Columbia River management and science prompted the Washington State Department of Ecology to request assistance from the National Research Council (NRC). The Department of Ecology contacted the NRC in 2002, and later that year the committee that authored this report was appointed. The NRC's Water Science and Technology Board, working in cooperation with the NRC's Board on Environmental Studies and Toxicology, appointed the ad hoc Committee on Water Resources Management, Instream Flows, and Salmon Survival in the Columbia River and coordinated the study. The committee conducted its deliberations and its report production in response to the Statement of Task given in Box 1-1. Consistent with the title of the study committee, and consistent with its statement of task, this report emphasizes the implications of water withdrawals from the mainstem Columbia River in the State of Washington (the "middle reach" of the Columbia) for Columbia River salmon. The proposed water extractions considered in this study have the potential to primarily alter two key physical characteristics in the impounded Columbia River as they affect salmon survival—water temperature and water velocity associated with river flow. These factors are of importance to salmonids migrating through the impounded Columbia. Additionally, one salmon species (ocean-type Chinook) spawns and rears in the mainstem Columbia River. This report deals generally with the issues of water withdrawals and instream flow conditions as they affect all Columbia River basin salmon species; but as federally-listed species are a special concern in the basin, parts of the report (e.g., a section of Chapter 4) focus on the effects of future environmental conditions on listed salmon species. Conditions in the river's tributary streams are also important to salmon survival rates, but given this report's focus on proposed mainstem water withdrawals, environmental conditions in tributary streams are only of peripheral interest in this study. The report also reviews and comments on several water management scenarios. (These scenarios were presented by the State of Washington and are listed in Appendix A.)

This committee held four meetings in 2003. Its first two meetings were in Richland, Washington, in February and in

> **BOX 1-1**
> **Committee on Water Resources Management, Instream Flows, and Salmon Survival in the Columbia River: Statement of Task**
>
> The committee will assess the risks to salmonids at critical stages in their life cycles under a range of different Columbia River system water management scenarios—including diversions for hydropower and other purposes—under both historical and present hydrological conditions.
>
> The study will:
>
> 1. Work with a science advisory panel (to be appointed by the Washington Department of Ecology) to gather information necessary to accomplish tasks 3 and 4, from the scientific community with direct experience in the Columbia River Basin, to include holding a workshop in Eastern Washington State.
>
> 2. Review and evaluate existing scientific data and analyses related to fish species listed under the Endangered Species Act in the Columbia River basin, as necessary to accomplish tasks 3 and 4.
>
> 3. Review and evaluate parameters critical to the survival and recovery of listed fish species as they relate to the hydrology of the Columbia River system in the context of the continued operation of the Federal Columbia River power system and other mainstem power generation facilities. This will include instream flows sufficient for fish and wildlife as well as the potential effects of decreased natural storage capacity on river hydrology.
>
> 4. In light of existing withdrawals, describe the risks to salmonid survival of a range of water withdrawals, and the cumulative effects of other factors, during critical times of the salmon life cycle (Note: the State of Washington Department of Ecology suggests an appropriate range of water withdrawals to consider is 250,000 acre-feet to 1,300,000 acre-feet).
>
> 5. Evaluate the effects of proposed management criteria, specific diversion quantities, and features of potential water management alternatives (such management information will be provided by the State of Washington).
>
> 6. Identify gaps in the knowledge and scientific information that are needed to develop comprehensive strategies for recovering and sustaining listed species and managing water resources to meet human needs.

Vancouver, Washington, in March, both of which included presentations from scientists from academia and federal and state agencies, representatives from regional stakeholder groups, basin water managers, and members of the public. At the first two meetings, members of a "Resources Group" (listed in Appendix B), convened by the Department of Ecology to provide scientific input to the study, provided several presentations on key scientific issues. The final two meetings were held in Olympia, Washington, in July and in Washington, D.C., in November, respectively, during which the committee discussed its Statement of Task and prepared its report.

In addition to the Resources Group experts, oral and written comments from many interest group representatives and the public were considered. In listening to and discussing comments from all presenters, it became clear that the issue of water withdrawal and management on the Columbia River is both a scientific and a public policy subject of regional as well as national importance. It was concluded that in order to comprehensively address the committee's task statement, agricultural, biological, cultural, economic, energy, environmental, historical, legal, and political factors all had to be considered. The more important challenge was thus not to decide whether or not to incorporate this diversity of knowledge into this report, but rather how to integrate it in a balanced manner that provided sound advice for managing water resources and salmon in the Columbia River system.

The challenges of managing Columbia River water and salmon defy simple solutions, and they are not likely to be successfully resolved with information from a single discipline or by the actions of a single group. Decision makers, scientists, policy analysts, and others must cooperate, as must entities across the basin. This report recommends some changes to water management processes in the basin. Successful implementation will require both cooperation and compromise. This is not to say that cooperation and compromise on Columbia water management issues has been absent, as there is a long history of government scientists working with policy makers on Columbia River water and salmon management issues. The Columbia River basin today, however, may be at a point where novel approaches to collaboration are in order. Humans and society have asked much from the Columbia River, and it has delivered a rich variety of benefits. But after several decades of human and technological interventions on the river and across the basin, the river system has fundamentally changed. In particular, salmon are at a critical point with regard to their long-term survival. If salmon habitat and populations are to be meaningfully protected and restored, people and organizations with stakes in Columbia Basin water may be required to make fundamental adjustments.

This report's organization reflects its multiple perspectives. Following this introductory chapter, Chapter 2 discusses the basin's broad physical, biological, and social features; Chapter 3 discusses hydrology and water management; Chapter 4 discusses environmental influences on salmon; Chapter 5 discusses laws and institutions; Chapter 6 discusses economics and water management alternatives; Chapter 7 discusses risks and water withdrawals; and Chapter 8 is a brief concluding epilogue. The target audience for this report is broad and includes science and policy experts, public- and private-sector officials, and individual citizens and stakeholder groups in the Columbia River basin in the western United States and Canada. This group includes Canadian and U.S. governors and legislators, tribal leaders, state-level water managers and staff (which includes the State of Washington Department of Ecology), federal agency staff (the Bonneville Power Administration, the Corps of Engineers, the Northwest Power and Conservation Council and its Independent Science Advisory Board, the Columbia Basin Project, NOAA Fisheries, and the U.S. Fish and Wildlife Service), other operators of dams and water diversion structures, Columbia River basin municipalities, farmers, commercial and recreational fishers, foresters, and tourism, recreational, and environmental organizations. Summaries are given at the end of each chapter. The report's principal conclusions and recommendations are printed in boldface in the Executive Summary and in Chapter 8.

2

Development and Changes in the Columbia River Basin

The Columbia River basin consists of several different physiographic regions. There are alpine and subalpine environments in its mountainous regions (the Cascades, Rockies, and related subchains), an arid and semiarid Columbia Plateau and other interior areas, and a more humid lower Columbia River valley. This breadth of physical regions is expressed in the basin's diversity of biomes, which include deserts, forests, shrubland, and riparian ecosystems. Much of the basin lies within the rain shadow of the Cascade Mountains and thus experiences an arid to semiarid climate. Precipitation is strongly seasonal; the majority of precipitation falls during the winter months, much of it as snow. The presettlement Columbia River experienced snowmelt-driven peak flows in May and June and lower flows in the fall. The Columbia's current flow patterns have been affected by a variety of human activities. Irrigated agriculture has diverted water from the Columbia and its tributaries. Logging has altered vegetative cover and landforms, which has in turn affected surface and groundwater flows. The nation's most extensive hydroelectric power system—the Federal Columbia River Power System (FCRPS)—was constructed on the Columbia River and its tributaries during the twentieth century. The system's numerous dams and storage reservoirs have altered both the volume and seasonal patterns of the Columbia's flows.[1] These changes to Columbia River discharge have affected the assemblage of fishes in the basin. With respect to their impacts on salmonid populations, some adverse changes have diminished in influence over time, while others have increased. Human-

---

[1] The system contains several run-of-the-river reservoirs that have minimal effects on river flows.

induced changes have interacted synergistically with certain natural factors, ameliorating some of these factors and exacerbating others. This chapter discusses key environmental and human features in the Columbia River basin and how human activities have impacted the basin's environmental systems. The basin's complex physical character and the changes induced by nineteenth- and twentieth-century agricultural, forestry, and industrial activities provide the context for considering more detailed aspects of changes to the Columbia River hydrological regime and its interactions with the life histories of Columbia River salmonids.

SETTLEMENT AND DEVELOPMENT OF THE COLUMBIA BASIN

Most inhabitants of the Pacific Northwest live in the Portland-Seattle urban corridor west of the Cascade Mountains (Portland lies within the Columbia River basin; Seattle does not). Of the roughly 9.5 million people in the four northwestern states (Idaho, Montana, Oregon, and Washington), about 5 million live in the Columbia River basin (Volkman, 1997). Like the rapidly growing Portland-Seattle corridor, the basin's interior has experienced population growth in many areas since the 1980s, with the largest increases in the urban areas of Bend (Oregon), Boise (Idaho), Richland/Pasco/Kennewick (the "Tri-Cities"), Spokane, Wenatchee, and Yakima (these last four urban areas are in Washington State). Beyond these cities, the rest of the basin is only sparsely populated.

Exploration and Settlement

Humans have inhabited the Pacific Northwest for at least 15,000 years (Jackson and Kimerling, 2003). Early inhabitants made a transition from hunting large game to a more sedentary lifestyle about 3,500 years ago, and salmon became an important part of their sustenance and culture. Even then, human activities affected salmon and salmon habitat. Native Americans who lived along and near the river expended considerable efforts in taking salmon from the river, and the populations of some riverside villages swelled during the peak of the salmon runs (White, 1995). Popular sites for catching returning salmon on their upstream journeys were at the Cascades and at Celilo Falls/The Dalles. Native Americans altered the landscape in their quest for salmon, with some consequent effects on the aquatic environment. European settlers introduced a new and more intensive set of harvesting techniques, which increased the scale and pace of environmental changes and attendant pressures on salmon stocks. European settlement in the region, and associated uses in resources and changes in the landscape, varied in timing and intensity across the Columbia River basin. This progression can generally be classified as follows: initial European settlement

Celilo Falls prior to the construction of The Dalles Dam. Photo courtesy of Ernest Smerdon.

(1810 to 1930s); mining, livestock, and agriculture (1850s to 1910); large-scale timber harvesting (1920 to 1990s); water diversions and mainstem dams (1900 to 1968; see Wissmar et al., 1994, for a review of the history of resource use in eastern Oregon and Washington).

The region's best-known and most celebrated exploration was the Lewis and Clark expedition. After traveling up the Missouri River and crossing over the Rocky Mountains, Lewis and Clark and their Corps of Discovery floated down the Columbia River to the Pacific Ocean, spending the winter of 1805-1806 at Fort Clatsop near present-day Astoria, Oregon. Lewis and Clark noted several characteristics of the streamside vegetation in their early-nineteenth-century exploration of the region, including an increase in riparian forests as one approached the ocean:

> The face of the country on both sides of the river, above and below the falls, is steep, rugged, and rocky, with a very small proportion of herbage, and no timber except a few bushes. (p. 669, referring to locations near The Dalles)
>
> Above Crusatte's river (Wind River) the low grounds are about three-quarters of a mile wide, rising gradually to the hills, with the rich soil covered with grass, fern, and other small undergrowth; but below the country rises with a steep ascent, and soon the mountains approach the river with steep rugged sides, covered with a very thick growth of pine, cedar, cottonwood, and oak. (p. 679, referring to farther downstream)
>
> At this village the river widens to nearly a mile in extent; the low grounds become wider, and they as well as the mountains on each side are covered with pine, spruce-pine, cottonwood, a species of ash, and some alder. After being so long accustomed to the dreary nakedness of the country above, the change is as grateful to the eye as it is useful in supplying us with fuel . . . the low grounds are extensive and well-supplied with wood . . . the low grounds near the river are covered so thickly with rushes, vines, and other small growth that they are almost impassable. (Coues, 1893, p. 668-691)

The Columbia's tributaries often had more abundant riparian vegetation than did the mainstem river:

> A branch of the Wollawollah river . . . is a bold, deep stream, about ten yards wide, and seems to be navigable for canoes. The hills of this creek are generally abrupt and rocky, but the narrow bottom is very fertile, and both possess 20 times as much timber as the Columbia itself. (ibid., pp. 978-979)

Other nineteenth-century explorers provided additional detail on the predevelopment vegetation and agricultural potential, describing the dominance of cottonwood and willow along inundated river banks from an elevation of 5,000 feet down to the river (Cooper, 1860). Although cottonwoods covered the islands and low shores of the lower Columbia River, upstream from The Dalles, willow and small hackberry were the only trees for hundreds of miles. The increasing scarcity of riparian vegetation as one moved eastward along the river corresponded with increasing aridity, a phenomenon observed by Lewis and Clark and railroad explorers and surveyors. Large tributaries of the Columbia apparently had a similar scarcity of timber. Cooper (1860) described the Yakima River as "wide, open, and destitute of timber, except in the bottom lands, and even there few trees are found for forty miles." The lower part of the Yakima basin was judged to be "less fit for cultivation than higher up, but contains much good grass land". Improvements in soil arability and in streamside timber that correlated with increasing elevation were emphasized: "On the immediate banks of the Columbia the country is not promising; but going back a little distance the grazing is very luxuriant and excellent, and the soil rich, particularly in the river valleys" (Cooper, 1860).

The basin is also notable for its variety of climatic regions and for sharp changes in climate zones over short distances. These contrasts near the Columbia River upstream and downstream from the Cascade Mountains were noted in the late nineteenth century:

Even from the Dalles we could perceive a thick fog hanging in the gap, but were quite unprepared to find a heavy rain, which we entered long before reaching the Cascades, and which continued unceasing during the whole day and night following, when we reached Vancouver. Even after entering this rain we could see the bright, unclouded sky of the plains eastward, but I thought the moister and milder air more agreeable than the cold dry climate we had just left. The change in the appearance of the country in the distance of a few miles was almost as great as I have since observed between New York and the isthmus of Panama in January, as we left the ground at the Dalles covered with snow, and entered a region of perpetual spring, with gigantic evergreen forests, tropical looking shrubs, and large ferns, where several spring flowers were still blooming. Even the perpendicular rocks supported a green covering of mosses, etc., over which cascades unbroken for a thousand feet, fell from the mountains directly into the river (ibid.).

Economic Activities and Sectors

Furs and Minerals

The British-controlled fur trade in north-central Washington began in the early 1800s. An active British and American fur trade, with furs being transported from the region to the mouth of the Columbia River, continued until midcentury. The decline of beavers and beaver dams reduced water storage in the uplands and reduced the environmental heterogeneity encountered by salmon. Discovery of gold in the 1850s attracted large numbers of miners to Washington and Oregon. For example, 1,200 to 3,000 miners mined the Similkameen River channel before moving north to the Fraser River in 1860, leaving behind a settlement near Oroville, Washington (Wissmar et al., 1994). The next 40 to 50 years saw numerous strikes of gold and silver and the appearance of boom towns in Washington. Placer and lode mining, mill wastes, and uncontrolled development degraded many sections of streams including Salmon Creek (Ruby City), which lost its large run of spring Chinook salmon.

Ranching and Irrigated Agriculture

Appreciable numbers of domestic stock were present in the basin by the 1860s. In the mid-1800s, settlers arrived via covered wagon and the Oregon Trail. The extension of railroads into the region in the late 1880s supported a subsequent and larger wave of settlers. Numbers of horses and livestock increased rapidly during the same period as well. Cattle from the Yakima and Willamette valleys supplied the northern mining camps. Cattle were abundant throughout the Yakima River valley by the 1870s. In the summers, cattle and sheep in large numbers were driven into headwater stream valleys. There were also large numbers of sheep in the John Day River basin near Shaniko, Oregon. For example, by 1904, Yakima County had 147,000 sheep, the largest number of any county in Washington (Wissmar et al., 1994). Between the 1850s and 1930s, overgrazing, deliberate burning to stimulate grass production, and wildfires increased soil erosion and sedimentation of streams. Remedies included restrictions on grazing in degraded areas, issuance of fewer grazing allotments, and lower allowable stocking rates.

Agriculture surpassed mining as the basin's principal economic activity in the early twentieth century. Although agricultural expansion was restricted by a lack of reliable water sources in many areas, some rudimentary irrigation canals were constructed as early as the 1850s. This stimulated settlement, and many cattle and sheep ranches sprang up across the basin, especially in the Yakima River valley. By 1869 a large irrigation canal watered lands below the confluence of the Naches and Yakima rivers, and many former grazing lands were converted to permanent, higher-value, horticultural crops. Passage of the Reclamation Act in 1902 and creation of the Reclamation Service (later renamed the Bureau of Reclamation) marked a new era in irrigated agriculture in the western United States. Today, much of the basin's agricultural production depends heavily on irrigation, and water diverted for agriculture is the largest off-stream water use in the Columbia system—over 6,500,000 acres, or 37 percent of total cropland in the area, is irrigated (Census of Agriculture, 1997). Over 93 percent of daily water use in the Columbia River basin (105,301 acre-feet per day) is for agriculture (ibid.). Irrigation typically uses water withdrawn from surface water supplies, while municipal supplies (domestic, commercial, and industrial) are typically from groundwater sources. More than one-third (37 percent) of farms in the basin have some irrigated acreage (ibid.). Most of the potatoes, sugar beets, hops, fruit, vegetables, and mint produced in the region are from irrigated land, as is a large portion of hay and grain production. Although the basin's economy is diversifying and growing, employment and per capita income in the area both remain below national averages. Agriculture and related services continue to be major employers in the basin, providing over 10 percent of employment. Farm owners, tenants, and ranch families represent 19 percent of households in the basin, compared to 2 percent nationally (Quigley et al., 1997). Within the agricultural sector, the cattle industry represents the largest share of agricultural income, accounting for approximately 29 percent of sales (ibid.).

The Columbia Basin Project is the region's largest irrigation project. Authorized by Congress in 1935, the project was developed in parallel with the construction of Grand Coulee Dam (which impounds Franklin D. Roosevelt Lake). Funds were allocated for construction of the dam in 1933, which was constructed to generate hydroelectric power as well as store irrigation water for the Columbia Basin Project (the Bureau of Reclamation constructed, and today manage, both the dam and the irrigation project). Initial designs of the project called for the delivery of irrigation water to 1,100,000 acres of land. About 671,000 acres are currently irrigated (*http://www.usbr.gov/dataweb/html/columbia.html*, last accessed December 5, 2003). The Columbia River Project stretches northward to the Canada-U.S. border and southward to Pasco, Washington. Crops raised on project lands include grains, alfalfa, hay, beans, fruit, sugar beets, potatoes, and sweet corn (ibid.; Chapter 3 discusses the Columbia Basin Project and its hydrological features in greater detail).

Forestry and Logging

Road construction facilitated logging and recreation in less accessible, higher-terrain areas. Greater access to riparian areas increased recreational activity, resulting in impacts to soil and vegetation near streams. Prior to road expansion, timber harvest and transport by water and horse was largely limited to lower-valley bottoms and adjacent slopes, with the timber used locally. Logging by truck in the Little Naches watershed began in 1931. Private land outside the Forest Reserve was completely logged by 1944. In 1975 the first timber clear-cuts appeared, and by 1992, 35 percent of the harvestable area of the watershed had been harvested. Timber harvesting and road construction in the upper Grand Ronde River basin have increased since the 1950s. Similarly, timber harvest is a dominant land use in the Blue Mountains (Ochoco, Umatilla, and Malheur national forests); the Blue Mountain Forest Reserve was established in 1906, and by the 1920s timber harvest was significant. Timber harvests across the basin steadily increased until about 1950, held constant through much of the 1990s, and have since decreased. Harvesting and grazing over the past century have reduced the tree canopy over many streams in the Columbia River watershed (e.g., in the John Day River basin, the entire canopy of many river sections has been removed). The environmental impacts of these actions include increased stream temperature, a reduction in areas of cold-water refugia for fish, and a reduction in ecologically beneficial inputs of coarse woody debris to the channel.

Many watersheds across the Columbia River basin are recovering from twentieth-century logging practices, such as splash dams, that had deleterious effects on streams. Changes in logging practices since the 1960s and 1970s, such as the addition of buffer areas, have helped reduce logging's impacts (e.g., soil erosion, slope failure). Similar trends are associated with the Columbia basin mining industry—many streams still show the effects of nineteenth- and twentieth-century activities. Impacts on the Columbia basin landscape from grazing and irrigated agriculture practices continue in much the same mode as they did through the twentieth century. Human population growth and its attendant effects, such as the paving of watersheds and pressures for additional water withdrawals for human activities, will be a major factor affecting Columbia River basin landscape and hydrological dynamics in the twenty-first century.

Human Population Projections

Human population in this region may reach 40 million to 100 million by the end of the twenty-first century. Estimates of population growth for the interior Columbia River basin to 2040 range from 0.3 percent per year (based on birth and death rates in the 1980s) to 1.6 percent per year (including immigration; McCool and Haynes, 1996). Nearly all of the basin's economic activities have affected Columbia River salmon and salmon habitat. The fact that so many human actions have affected salmon habitat in so many different ways confounds scientific investigations of the relative impacts of a given activity(ies). Yet the fact that the region's human population seems highly likely to continue growing (with substantial growth in some regions in or near the basin, especially the Portland-Seattle corridor) under current population and immigration policies suggests that pressures for water and related services (e.g., hydroelectricity) will likewise continue to grow, which will exert more pressure for additional diversions of water from the Columbia River mainstem and tributaries. As a previous National Research Council committee that reviewed Columbia River salmon management stated, "As long as human populations and economic activities continue to increase, so will the challenge of successfully solving the salmon problem" (NRC, 1996).

FEDERAL COLUMBIA RIVER HYDROPOWER SYSTEM

The FCRPS was constructed and is managed and operated by three federal agencies: the Bonneville Power Administration (BPA), the Bureau of Reclamation, and the Corps of Engineers. The system consists of 31 dams on the Columbia River and its tributaries and the related power generation and transmission infrastructure. The system's dams and reservoirs impound roughly 55,000,000 acre-feet of water (FCRPS, 2001). It has a generating capacity of roughly 33,000 megawatts (NPCC, 2004) and provides about 60 percent of the region's hydroelectricity generating capacity (FCRPS, 2001). This system provides the Pacific Northwest with the lowest power rates in the nation and has been an important factor in attracting industries such as aluminum smelting and aircraft manufacturing. In addition to the economic and social changes that resulted from the project's construction and operations, it also fundamentally restructured the Columbia's hydrological character and its related ecological resources.

Plans for the systems' construction were under way in the early twentieth century. In the mid-1920s, Congress requested the Corps of Engineers to conduct a survey of the basin's potential for the construction of dams and related works to promote hydroelectricity production, irrigation, flood control, and navigation. In 1931 the Corps issued a comprehensive study of the Columbia and its prospects for multipurpose development (the document was part of the Corps "308 reports," so named after the U.S. House Document authorizing them and which were conducted for several major U.S. river basins). The Corps of Engineers 308 report called for 10 dams on the Columbia, and the report shaped the river's development for the next 40 years (FCRPS, 2001).

The early 1930s were a period of technological optimism, with a strong faith in the ability of multipurpose river basin development to deliver substantial social and economic benefits. The federal Tennessee Valley Authority was established in 1933, and presidential candidate Franklin Roosevelt promised hydroelectric development of the Columbia River while campaigning in Portland, Oregon, in 1932. Many saw electrification of the Columbia as central to the region's development and as an antidote to the Great Depression's economic woes. Construction began on both Bonneville and Grand Coulee dams in 1933. In 1937 the Bonneville Project Act was signed, which created the BPA to market power from the two dams. The agency was mandated to construct and operate transmission facilities and market hydroelectricity, while responsibility for dam operations remained with the Corps and the U.S. Bureau of Reclamation. In 1939, BPA first transmitted energy from Bonneville Dam to Cascade Locks, Oregon, and then later to Portland. Grand Coulee Dam first provided power to the BPA system in 1941. The BPA and the FCRPS have since played crucial roles in the region's economic development. The large supply of low-cost power provided by the FCRPS enticed many industries to locate in the region, most notably aluminum smelting and aircraft production. Boeing Aircraft Works in Seattle ramped up production in the World War II era, and other wartime industries followed. BPA and the FCRPS were vital to World War II industrial production, as BPA also marketed power to the Hanford Reservation for plutonium production. BPA marketed power produced from the Hanford Generating Plant, which was part of the Washington Public Power Supply System. The BPA has also been an important participant within the processes of the Northwest Power and Conservation Council (known until 2003 as the Northwest Power Planning Council, or NWPPC).

The Northwest Power Planning Council (today's NPPC) was created in connection with passage of the Pacific Northwest Electric Power Planning and Conservation Act of 1980 (P.L. 96-501). The council was formed with representatives from the states of Idaho, Montana, Oregon, and Washington. The act directed the council to draft a plan for meeting the region's electrical needs at the lowest cost. The council was also charged with developing a fish and wildlife program (in addition to a power plant) that directs the Bonneville Power Administration to fund projects to enhance fish and wildlife resources (at $100 million to 150 million per year; see *http://www.nwcouncil.org*, last accessed March 3, 2004). The BPA was given responsibility to meet electrical demand while managing the system to meet the act's purposes relating to fish, system efficiency, and experimental projects (available online at *http://www.nwppc.org/library/2003/2003-2.pdf*, last accessed December 5, 2003). The act's emphasis on equitable treatment of fish and wildlife drove efforts to rebalance FCRPS operations during the 1980s and 1990s. Key guidance in operating the system to provide instream flows and help protect endangered fish species has been provided in

Biological Opinions issued by NOAA Fisheries (formerly the National Marine Fisheries Service). These documents are issued in response to Biological Assessments submitted by federal action agencies (e.g., Bureau of Reclamation) pursuant to the Endangered Species Act.

Beyond system operations, construction of the hydropower system itself also had notable environmental consequences. The reservoirs inundated and eliminated almost all mainstem spawning areas, with the exception of the Hanford reach (a stretch of river downstream of the federal Hanford nuclear facility in central Washington). The Grand Coulee (Columbia River) and Hells Canyon dams (three dams on the Snake River) blocked large amounts of habitat that were once highly productive salmon habitat in the Columbia basin. The tributary habitat that today produces spring Chinook and steelhead is the fringe habitat that remains (Dauble et al., 2003). The dams also inundated vast acreages of wildlife habitat.

SUMMARY

Human activities have long had significant impacts on Columbia River salmon and aquatic habitat. Activities of Native Americans impacted the salmon, as tribal actions altered the landscape and affected the aquatic environment. But the introduction of industrial-based economic activities to the Columbia River basin, and the consequent settlement and human population growth, resulted in widespread and substantial changes in land uses and basin hydrology. The basin has been developed and altered by additional settlement and population growth, extractive activities (e.g., mining and trapping), agriculture and ranching, large-scale timber harvesting, water diversions, and mainstem dams and reservoirs. The basin's economy has historically depended heavily on the Columbia River, first through the harvest of salmon, and then later through the construction of dams and related infrastructure to promote irrigated agriculture and hydroelectric power development, to provide flood control, and to support navigation. The Columbia River has clearly yielded a wealth of benefits to the region and its inhabitants. But the impacts of these various activities have had substantial effects on Columbia River salmon. As a result of these activities, some salmon runs have gone extinct and several of the basin's anadromous salmon are today listed as threatened and endangered under the Endangered Species Act. Over the years, regional and federal water and fisheries management organizations have enacted several strategies designed to mitigate environmental impacts on the salmon. As this chapter has discussed, these impacts have derived from several different activities. Strategies aimed at replacing natural ecosystem processes that have been lost or compromised cover a wide spectrum of practices, including fish ladders, the transporting of salmon around dams, and dam operations (NRC, 1996; the next chapter discusses "flow targets," or instream flows designed to meet the needs of salmon). Other strategies could include changes in human uses of tributary riparian systems, changes in logging practices and policies, hatchery management practices, or changes in ocean salmon harvest policies. The point is that salmon have been affected by a wide variety of human activities and that policies designed for protecting or enhancing salmon populations may need to assume a similar breadth. The potential additional water withdrawals from the Columbia River considered in this study thus make up only a portion of a large and complex mosaic of human activities that affect salmon.

The human population in the interior Columbia Basin in the United States is about 5 million and projected to grow by 0.3 to 1.6 percent per year. Human population growth has implications for salmon survival, not only because of urbanization's direct effects on land use and hydrology (e.g., changing of timing of runoff patterns, decreasing of surface waters percolating to groundwater) but also because additional people will generate additional demands for Columbia River water and related resources. The region has changed dramatically over the past 150 years and given human population growth projections, even more rapid future changes are likely. As discussed in this chapter, construction of the FCRPS resulted in marked and lasting changes to the basin's physical and economic systems. The following chapter examines details regarding construction of the system for Columbia River basin hydrology as well as other important hydrological changes wrought by decades of human activities in the region.

3

Hydrology and Water Management

Columbia River water flows have generated enormous social and economic benefits. These uses include hydropower generation, flood control, instream flows for fish and habitat, irrigation, commercial and subsistence fishing, navigation, and water for municipalities and industries. A vast number of jurisdictions and individuals use Columbia River water, including seven U.S. states, the Canadian province of British Columbia, and several Indian reservations (Figure 3-1). The geographical focus in this study, however, is on the mainstem Columbia River in the State of Washington.

August 1955. Fishing at Celilo Falls, where The Dalles dam was later constructed. In the background, an American Indian prepares to take a fish out of the net from a scaffold. Photo courtesy of Ernest Smerdon.

Hydrology and Water Management 43

As explained in Chapter 2, there are many large dams (storage and run-of-the-river) and reservoirs along the river that compose the Federal Columbia River Power System (FCRPS). The Columbia River dams in Washington State are owned and operated by federal entities and by state public utility districts. Their daily operations are designed to meet the needs of many sectors, the most important being flood control, hydroelectric power generation, and instream flows. Like most regions of the western United States, irrigated agriculture is the largest consumptive water user in the region. Irrigated agriculture along the Columbia River in the State of Washington consists of one very large withdrawal—the Columbia Basin Project—and a large number of small (relative to the Columbia's flows) withdrawals by individual irrigators. These structures and uses have affected stream flows, water quality, and water temperature. This chapter examines twentieth-century changes in Columbia basin hydrology and the annual hydrograph, the current and prospective future picture of water withdrawals (this study's primary focus), water quality, and changes in water temperature and related prospective future changes in basin climate.[1]

This study focuses on the implications of water withdrawals from the mainstem Columbia River for salmon survival. An analysis of the relative impacts of mainstem surface and groundwater withdrawals in comparison to the hydrological impacts of Columbia River dam and reservoir construction and operations was beyond the scope of this study. This report focuses on mainstem water withdrawals because this topic was central to the committee's Statement of Tasks, not because of the relative

---

[1] This chapter includes several figures and tables containing hydrologic information. Some of those data are expressed in cubic feet per second (cfs) and some of the data are expressed in acre-feet per year (AF/yr.). This report does not present all hydrologic data in a single unit because both units (cfs and AF/yr) are traditionally and currently used by water managers, farmers, and scientists in different settings in Washington and across the western U.S. Furthermore, cfs represents a rate, while acre-feet represents a volumetric measure. For comparative purposes, however, 1 cubic foot/second of water equates to slightly less than 2 acre-feet/day—or roughly 724 acre-feet/year.

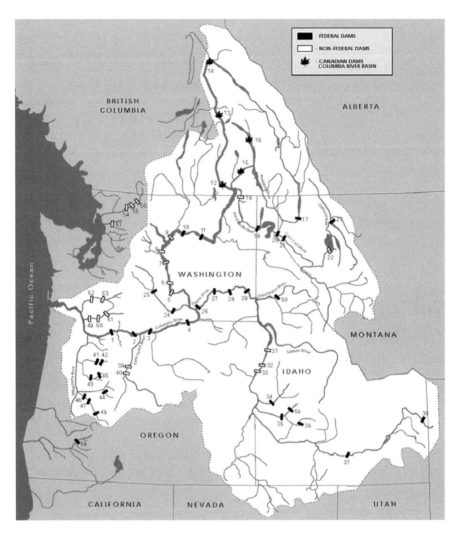

FIGURE 3-1 Columbia River basin and federal and non-federal dams.
SOURCE: FCRPS (2001).

MAJOR NORTHWEST DAMS		
1. BONNEVILLE Columbia River, USACE	21. NOXON, RAPIDS Clark Fork River, WWP	41. BIG CLIFF N. Santiam River, USACE
2. THE DALLES Columbia River, USACE	22. KERR Flathead River, MPC	42. DETROIT N. Santiam River, USACE
3. JOHN DAY Columbia River, USACE	23. HUNGRY HORSE Flathead River, USBR	43. FOSTER S. Santiam River, USACE
4. McNARY Columbia River, USACE	24. CHANDLER Yakima River, USBR	44. COUGAR McKenzie River, USACE
5. PRIEST RAPIDS Columbia River, Grant Co. PUD	25. ROZA Yakima River, USBR	45. GREEN PETER M. Santiam River, USACE
6. WANAPUM Columbia River, Grant Co. PUD	26. ICE HARBOR Snake River, USACE	46. DEXTER Willamette River, USACE
7. ROCK ISLAND Columbia River, Chelan Co. PUD	27. LOWER MONUMENTAL Snake River, USACE	47. LOOKOUT POINT Willamette River, USACE
8. ROCKY REACH Columbia River, Chelan Co. PUD	28. LITTLE GOOSE Snake River, USACE	48. HILLS CREEK Willamette River, USACE
9. WELLS Columbia River, Douglas Co. PUD	29. LOWER GRANITE Snake River, USACE	49. MERWIN Lewis River, PP&L
10. CHIEF JOSEPH Columbia River, USACE	30. DWORSHAK N.F. Clearwater River, USACE	50. YALE Lewis River, PP&L
11. GRAND COULEE Columbia River, USBR	31. HELLS CANYON Snake River, IP	51. SWIFT Lewis River, PP&L
12. KEENLEYSIDE Columbia River, BC Hydro	32. OXBOW Snake River, IP	52. MAYFIELD Cowlitz River, TCL
13. REVELSTOKE Columbia River, BC Hydro	33. BROWNLEE Snake River, IP	53. MOSSYROCK Cowlitz River, TCL
14. MICA Columbia River, BC Hydro	34. BLACK CANYON Payette River, USBR	54. GORGE Skagit River, SCL
15. CORRA LINN Kootenay River, W. Kootenay	35. BOISE RIVER DIVERSION Boise River, USBR	55. DIABLO Skagit River, SCL
16. DUNCAN Duncan River, BC Hydro	36. ANDERSON RANCH Boise River, USBR	56. ROSS Skagit River, SCL
17. LIBBY Kootenai River, USACE	37. MINIDOKA Snake River, USBR	57. CULMBACK Sultan River, Snohomish Co. PUD
18. BOUNDARY Pend Oreille River, SCL	38. PALISADES Snake River, USBR	58. LOST CREEK Rogue River, USACE
19. ALBENI FALLS Pend Oreille River, USACE	39. PELTON Deschutes River, PGE	59. LUCKY PEAK Boise River, USACE
20. CABINET GORGE Clark Fork River, WWP	40. ROUND BUTTE Deschutes River, PGE	60. GREEN SPRINGS Emigrant Creek, USBR

influence of withdrawals in comparison to other system users or management objectives.

COLUMBIA RIVER FLOWS

Changes to the Hydrograph

The annual Columbia River hydrograph underwent fundamental changes during the twentieth century. These changes were driven primarily by the construction of dozens of dams and reservoirs on the river's mainstem, hundreds of projects on tributary streams (some of these dams, such as those on the Snake River, are also quite large), and system operations. Although constructed to serve multiple purposes, the driving force behind Columbia River dam construction was hydroelectric power development and, to a lesser extent, flood control. With its solid rock channel, low levels of silt, and relative steepness, the Columbia River was uniquely suited for large-scale hydropower development.

Construction of the first federal Columbia mainstem projects began in 1933 at Bonneville and Grand Coulee. World War II increased pressure to further tap the river's hydroelectric power production potential, and between 1944 and 1945 Congress authorized several water projects in the basin. In the five years following the war, Chief Joseph Dam, Albeni Falls, Libby, John Day, and The Dalles dams were all authorized (Volkman, 1997). Support for federal dams on the mid-Columbia faded during the 1950s, but licenses were issued to county public utility districts to construct Priest Rapids Dam, Rocky Reach Dam, Wanapum Dam, and Wells Dam, all of which today are operated by public utility districts. Upstream dams that augmented storage and power production capabilities were constructed pursuant to the Columbia River Treaty signed between Canada and the United States in 1961; these dams included Libby Dam in Montana and Arrow Lakes, Duncan, and Mica dams in Canada. The treaty focused primarily on two water management sectors: hydropower and flood control.

The hydrological implications of the construction of this dam and reservoir system, and the operations of that system, on the Columbia River annual hydrograph were tremendous. Figures 3-2 and 3-3 provide two different portrayals of these changes. The long-term (1879-2002) average discharge of the Columbia River, as measured at The Dalles, is roughly 139,000,000 acre-feet per year (or roughly 192,000 cfs; USGS, 2003). Figure 3-2 shows how annual Columbia River hydrological seasonality has "flattened" the river's annual discharge patterns, as original high seasonal ("summer") flows have decreased and low seasonal ("winter") flows have increased. Figure 3-3 shows how the balance of flows between summer (April-September) and winter (October-March) has changed since the late 1800s. Through time this summer-winter division of flows has become closer to a 50:50 balance in response to system construction and operations largely oriented to serve hydroelectric power needs and operations. In addition to the smoothing of the annual Columbia River hydrograph, construction and operations of the dam and reservoir system have had two other major physical impacts: water velocities have decreased (which has significantly increased the amount of time required for juvenile salmon to travel downstream and into the sea), and the size and orientation of the Columbia River plume (a zone of fresh water extend-

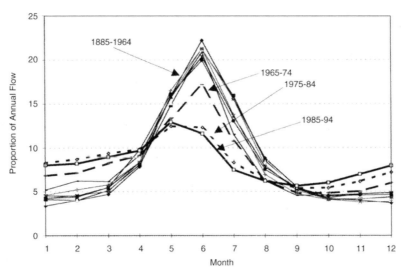

FIGURE 3-2 Distribution of monthly flows at The Dalles by 10-year blocks. SOURCE: Volkman (1997).

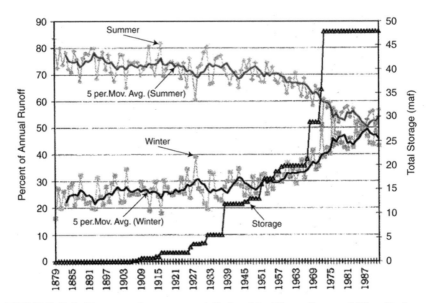

FIGURE 3-3 Changes in seasonal Columbia River flows at The Dalles, 1879 to 1992. SOURCE: Volkman (1997).

ing from the mouth of the Columbia into the Pacific have been greatly altered (Ebbesmeyer and Tangborn, 1992). This is particularly an issue when salmon smolts (young salmon two to three years old) are moving downstream. These changes, however, have not eliminated all variability of Columbia River flows. Figure 3-4, for example, demonstrates that considerable variability of annual Columbia River discharge exists between years. Flows also continue to vary on other timescales; for example, daily flow patterns below hydropower dams often vary substantially as flows are adjusted to demands in electric power. The cumulative impact of all these hydrological changes has likely had significant effects on the early ocean survival of juvenile fish leaving the Columbia River (Pearcy, 1992). Because of concerns over possible impacts on salmon from the construction and operation of the hydropower system, federal and state management and resources agencies have implemented some changes to system operations to help provide instream flows designed to support and enhance salmon habitat. The flows are referred to as "flow targets" and are discussed in the following section.

Hydrology and Water Management

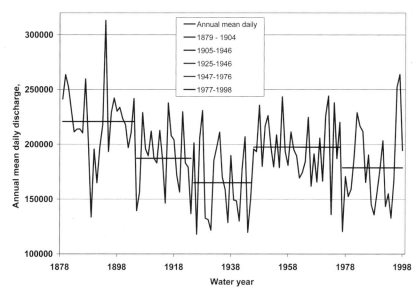

FIGURE 3-4 Columbia River discharge (in cfs), 1879 to 2000, at The Dalles, Oregon. SOURCE: USGS (2003).

Biological Flow Targets

Passage of federal environmental legislation in the 1960s and 1970s, such as the Environmental Policy Act (1969) and the Endangered Species Act (1973), led to changes in system operations as some flows were devoted to protect and sustain endangered salmon species and habitat. "Flow targets" were developed by federal and state resources agencies in efforts to ensure adequate instream flows. Key flow targets involving fishery resources include consideration of smolt migration, spawning flows for chum salmon below Bonneville Dam, spawning and incubation flows at Vernita Bar (see Box 3-1), water elevations in storage reservoirs, and minimum instream flows at reservoir outlets. Specifications regarding these flow targets are provided in the 2000 National Marine Fisheries Service Federal Columbia River Power System Biological Opinion (NMFS, 2000). Of all these flow targets, the most critical with respect to this discussion involves smolt migration flows.

> **BOX 3-1**
> **Venita Bar Agreement**
>
> The Vernita Bar Settlement Agreement, approved in 1988, ensures flows to incubate fall Chinook embryos and fry at Vernita Bar, a large gravel bar and an important spawning area 4 miles downstream from Priest Rapids Dam. Signatories to the agreement include the Bonneville Power Administration (BPA), the National Marine Fisheries Service (since renamed NOAA Fisheries), the Confederated Tribes of the Umatilla Indian Reservation and Colville Indian Reservation, the Yakama Indian Nation, the Washington Department of Fisheries (now Washington Department of Fish and Wildlife), the Oregon Department of Fish and Wildlife, and the Grant, Chelan, and Douglas county public utility districts.
>
> The agreement specifies how the Grant, Chelan, and Douglas public utility districts and BPA will provide the required flows and identifies special conditions in the event of inability to perform and adverse water conditions. Flows are regulated to minimize excavation of salmon redds (spawning nests) at flows higher than 70,000 cubic feet per second (kcfs). Grant County Public Utility District is to operate the Priest Rapids Project to the extent feasible to yield river flows during daylight equal to 68 percent of daily mean inflow to Wanapum pool. The agreement does

As pointed out, the Columbia River system continues to exhibit significant variability of discharge on many different timescales. Partly as a result of this variability, migration flow targets are not always met, and it has generally proven difficult to maintain mainstem flows above the target for the entire fish migration period. In years of low to moderate precipitation, decreased flows in the Columbia River exacerbate this phenomenon. Furthermore, because of consumptive use and hydropower demands during low-flow years, tradeoffs between fishery demands often come into play, particularly between biological needs within storage reservoirs and the associated outlets and anadromous migration conditions in the mainstem. Competing biological demands for water thus often make it impossible to achieve stated flow targets. Although these target flows have at times not been met, meeting the needs of biological and ecological objectives has become an objective with operational priority on par with flood control and hydroelectric power generation.

not obligate BPA to limit fall discharge, but BPA attempts to do so. After the end of the salmon spawning season, a field inspection assesses the protection level flow (minimum flow to protect established salmon redds) by several criteria. Protection of redds is related to flow levels in the guidelines. The protection level even considers details such as weekdays versus holidays or weekends and is highly specific. Some flexibility is permitted within the foregoing schedule as long as alternatives provide an equivalent volume. The biological monitoring program tracks temperature data to predict dates of spawning, hatching, emergence of fry, and the end of emergence. At the end of emergence, usually in mid-May, the protection flow level is terminated.

As this report went to press, an expanded Vernita Bar Agreement was being drafted. The new agreement, tentatively called the "Hanford Reach Fall Chinook Protection Program," is to be submitted to the Federal Energy Regulatory Commission for consideration in the relicensing process.

Hydropower Generation and Spill

The FCRPS is operated to furnish electrical power for industrial, urban, and agricultural needs. This results in daily variability in discharge to meet high demand during daylight hours and low demand during nighttime hours. Coordination between release of water from one dam to the next is important because (1) there are relatively short distances between the dams; (2) most of the dams are run-of-the-river, with little or no storage; and (3) the outflow from one dam is usually the start of the pool behind the next downstream dam. There is also the need to allow some "spill" (the bypassing of water around hydropower generation turbines) of water during downstream migration of salmon smolts to enhance their prospects for survival. The fish spill program is implemented during both the spring and summer smolt migration periods, from April through August. This strategy is designed to intentionally discharge water over spillways at different dams in the FCRPS in accordance with guidelines specified in NOAA Fisheries 2000 Biological Opinion. These operational features result in sharp changes in diurnal discharge patterns at dams on the Columbia and Snake rivers.

WATER WITHDRAWALS

Existing Water Rights

The Washington State Department of Ecology issues water rights permits in the state. After water use has commenced, department staff visit the site of use and issue a certificate. State law specifies that, if the full volume of a water right is not used at its allocated rate over a five-year period, the volume of the water right not used can be taken away. Historical trends indicate that most permit holders do not divert their full allocations during most years.

The Washington State Department of Ecology has, to date, issued 754 permits for surface water withdrawals from the mainstem Columbia River between the Canadian boundary and Bonneville Dam. The total maximum withdrawal volume of these permits is 4,240,000 acre-feet per year. Withdrawal permits held by the Columbia Basin Project total 3,160,000 acre-feet per year, which represents 74 percent (by volume) of the water rights issued in this reach of the Columbia River. The department has also issued 110 water rights for groundwater extractions within 1 mile of the River, which amount to 440,000 acre-feet per year. Permits in the State of Washington currently issued for Columbia River surface water and groundwater withdrawals within 1 mile of the river thus amount to about 4,700,000 acre-feet per year. An itemized list of surface water permits showed that roughly 96 percent of surface water diversions were used for irrigation, with the remaining 4 percent being used by municipalities and other uses (figures based on data provided by John Covert, Washington State Department of Ecology, 2003).

Table 3-1 illustrates and compares permitted volumes of water withdrawals from the mainstem Columbia River and from groundwater within 1 mile of the river, with regard to maximum,

TABLE 3-1 Columbia River Flows at John Day Dam, 1960-1999 and monthly Columbia River withdrawals

Month	(1) Maximum	(2) Mean	(3) Minimum	(4) Withdrawals	(5) percent of max.	(6) percent of mean	(7) percent of min.
Jan	16,200	9,690	5,430	10.8	0.1	0.1	0.2
Feb	18,200	9,500	5,740	10.0	0.1	0.1	0.2
Mch	20,400	11,100	6,200	110	0.5	1.0	1.8
April	19,800	12,100	5,920	597	3.0	4.9	10.1
May	29,400	17,200	8,110	765	2.6	4.5	9.4
June	34,700	19,000	7,120	792	2.3	4.2	11.1
July	21,400	12,500	5,110	850	4.0	6.8	16.6
Aug	13,400	8,390	5,420	793	5.9	9.5	14.6
Sep	9,260	6,420	4,280	498	5.4	7.8	11.6
Oct	10,400	6,910	5,430	274	2.6	4.0	5.1
Nov	9,280	7,340	5,170	12.3	0.1	0.2	0.2
Dec	15,100	8,870	5,210	11.7	0.1	0.1	0.2

Notes:
Columns 1-3—Maximum, mean, and minimum monthly discharges for Columbia River at John Day Dam. Values in thousands of acre-feet/month.
Column 4—Permitted volumes from mainstem Columbia River surface water withdrawals and groundwater from within one mile of the river, between the Canada-U.S. border and Bonneville Dam. Values in thousands of acre-feet/month.
Columns 5-7—Withdrawals as percentages of monthly Columbia River discharge values at John Day Dam.
SOURCE: USGS, 1996; Washington Department of Ecology, 2003.

mean, and minimum monthly discharges at John Day Dam[2] (1960 to 1999 database, USGS, 2001). Columns 1 through 3 list 1960-1999 Columbia River discharge figures recorded at John Day Dam. Column 4 lists the monthly distribution of water withdrawal permits along the river in the State of Washington. These monthly values are based on actual monthly withdrawal data at Franklin D. Roosevelt Lake by the Columbia Basin Project (agriculture, 96 percent of the withdrawals) and the city of Pasco (municipalities and industry, the remaining 4 percent). No water use data on groundwater withdrawals were available, so Table 3-1 assumes that 75 percent of groundwater withdrawals were used for irrigation and that 25 percent of groundwater was used for commercial, industrial, municipal, domestic, and other uses.

Consumptive use at the Columbia Basin Project is about 70 percent of the volume of surface water withdrawals (Montgomery Water Group, 1997; see also Appendix C). This 70 percent figure was assumed to apply to other areas of irrigated agriculture along the Columbia River mainstem (keeping in mind as well that the Columbia Basin Project represents the largest irrigated agriculture diversion along the river), with the remaining 30 percent of withdrawals eventually returning to the Columbia River as irrigation return flows and groundwater seepage. With regard to the municipal and industrial (M&I) water reflected in Table 3-1, data indicate that roughly 30 percent of municipal water is returned to the Columbia River through wastewater treatment plants (City of Pasco, 2003). It was further assumed that an additional 10 percent of M&I withdrawals returned to the Columbia River through groundwater seepage, for a total consumption of 60 percent of M&I water withdrawals.

Column 5 in Table 3-1 shows that withdrawals of existing water permits under high-flow conditions, as a percentage of total flows, ranged from 0.1 (in December) to 5.9 percent (in August). In contrast, Column 7 shows that withdrawals under

---

[2] John Day Dam was used as a reference site because almost all existing Columbia River consumptive withdrawals are upstream of this dam. Nearly all the pending permits for additional consumptive withdrawals in the State of Washington are also upstream of John Day Dam. Columbia River discharge figures at John Day include inflows from the Snake River; discharge data in Table 3.1 are thus higher than they would be for stations upstream of the Columbia-Snake confluence.

minimum flow conditions ranged from 0.2 (in January) to 16.6 percent (July). The critical months of withdrawals under minimum flow conditions are in July and August. These months are periods of high water withdrawals for irrigated agriculture and municipalities. The pronounced seasonality of withdrawals and the sharp differences in the effects of withdrawals according to season are key messages from Table 3-1. These data show that January withdrawals have very little effect on the overall flows of the Columbia but that during July and August current withdrawal volumes have noticeable effects on mainstem flows, especially during lower-than-average discharge years.

Columbia Basin Project

The U.S. Bureau of Reclamation's Columbia Basin Project (Figure 3-5) is the largest irrigation project in the Columbia River basin. The centerpiece of the project is the Grand Coulee Dam. Completed in 1941, it is the nation's largest concrete dam. It impounds about 9,400,000 acre-feet of water in Franklin D. Roosevelt Lake, which provides water to the East Columbia Basin Irrigation District, the Quincy Columbia Basin Irrigation District, and the South Columbia River Basin Irrigation District (*http://www.usbr.gov/dataweb/html/columbia.html*, accessed February 5, 2004). The most important crops on the project are alfalfa, apples, corn, potatoes, and wheat. The project's network of canals, tunnels, reservoir and pumping plants was intended to deliver water to about 1,100,000 acres of irrigated farmland, but today about 671,000 acres are irrigated (ibid.). Irrigation return flows from the Columbia Basin Project are discharged into the Columbia River through wasteways, creeks, and groundwater seepage.

Withdrawals

Rates and patterns of withdrawals at the Columbia Basin Project vary within and between years. Table 3-2 displays average monthly pumping rates from Franklin D. Roosevelt Lake

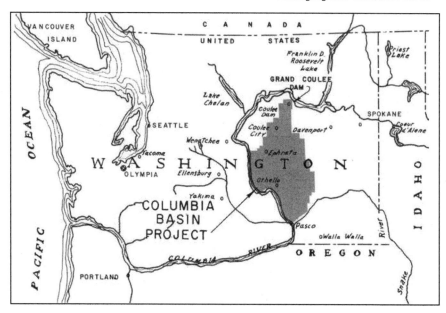

FIGURE 3-5 The Columbia Basin Project. SOURCE: Courtesy of U.S. Bureau of Reclamation, Ephrata, Washington.

(from which water is conveyed to the Columbia Basin Project) and shows that irrigation water is generally applied from March through October, with highest usage during June and July. Figure 3-6 shows annual withdrawals for 1975 to 2000 from Franklin Roosevelt Lake to the Columbia Basin Project. Maximum and minimum annual values were 3,090,000 acre-feet and 1,450,000 acre- feet per year, respectively. The low withdrawal in 1980 corresponds to the eruption of Mount St. Helens. Note that in only one year—1995—did project withdrawals approach the permitted maximum (the U.S. Bureau of Reclamation has water rights for 3,158,000 acre-feet of water per year at Grand Coulee). Expansion of irrigated agriculture on Columbia Basin Project lands would increase withdrawals toward this permitted maximum, which would reduce downstream flows (although roughly 30 percent of the additional withdrawals would return to the Columbia River, a figure that could decrease over time with more efficient irrigation systems). During the time period displayed in Figure 3-6, annual withdrawals averaged roughly 80 percent of the permitted level (1990 withdrawals, depicted in

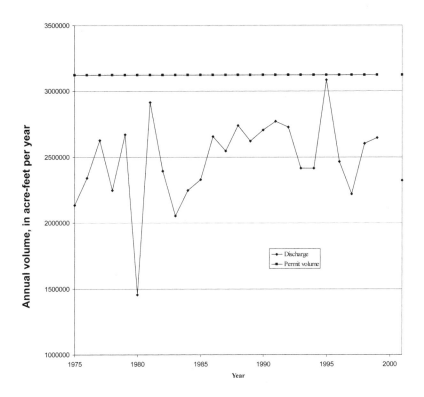

FIGURE 3-6 Annual withdrawals from the Columbia River at Grande Coulee Dam by the Columbia Basin Project. SOURCE: USGS (2003a).

Table 3-2, represented an above-average annual withdrawal for the time period in Figure 3-6). If the Columbia Basin Project were to withdraw its full entitlement of water each year, it would result in an average 25 percent increase of water being delivered to the project (an increase in withdrawals of about 600,000 acre-feet per year). As mentioned, the Columbia Basin Project currently accounts for roughly 74 percent of total water withdrawals from the middle reach of the Columbia River in the State of Washington.

Potential Additional Withdrawals from the Columbia River

One focus of this study was consideration of the effects and risks to salmonid survival over a specific range of proposed additional water withdrawals (250,000 acre-feet per year—1,300,000 acre-feet per year). An additional 1,300,000 acre-feet per year of water withdrawals from the mainstem Columbia River from the Canada-U.S. border to Bonneville Dam would constitute roughly a 28 percent increase in the volume of water permits that have been issued by the State of Washington for surface water withdrawals from the Columbia River and groundwater withdrawals from within 1 mile of the river (current permitted total maximum volume is 4,240,000 acre-feet per year). The effects of these proposed additional withdrawals, and their attendant risks, will vary considerably depending on flow levels in the Columbia River at any given time. Under current withdrawal patterns, the greatest effects of withdrawals on flows are during July and August (particularly during low-flow years), as these are the months of highest withdrawals. The seasonality of proposed additional withdrawals was assumed to be similar to existing water uses for irrigation and municipal uses and is of overriding importance in considering the implications of Columbia River withdrawals for salmon survival. Under current conditions, during January, less than 1 percent of total annual withdrawals are made. About 18 percent of total annual withdrawals are made in July.

TABLE 3-2 Average monthly volumes (thousands of acre-feet) of water pumped from Franklin D. Roosevelt Lake, 1990.

Month	Mar	April	May	June	July
Volume	38	387	508	525	539
Percent of Total Annual Withdrawal	1.3	13.4	17.6	18.2	18.7

SOURCE: Bonneville Power Administration (1993).

Hydrology and Water Management 59

Numerous calculations and speculations could be made regarding to the proposed range of withdrawals. Assuming that the seasonal pattern of withdrawals continues essentially unchanged and that the upper end of the range of proposed additional withdrawals (1.3 million acre-feet per year) is diverted, this would entail additional withdrawals of roughly 2,600 acre-feet in January and roughly 234,000 acre-feet in July. The effects in January of the upper end of the proposed range of additional diversions would still result in total withdrawals being less than 1 percent of mean January Columbia River flows. The effects in July of the upper end of the proposed range of diversions, by contrast, would increase July withdrawals from roughly 6.8 percent of mean Columbia River flows to roughly 8.6 percent of mean Columbia River flows at John Day Dam (based on 1960 to 1999 flows; see Table 3-1). Under *minimum* July flow conditions, the upper end of the proposed range of diversions would increase July withdrawals from roughly 16.6 percent of Columbia River *minimum* flows to roughly 21 percent of Columbia River flows at John Day Dam.

In addition to permit applications for withdrawals currently being considered by the State of Washington, other factors point to the possibility of further reductions in future Columbia River flows. Regional climate warming could reduce flows in low-flow periods, human population growth is likely to exert pressures for additional withdrawals from the Columbia, and current users (e.g., tribal reservations) may seek to increase current levels of withdrawals. The occasional but virtually certain coinci-

August	September	October	Annual	Permitted Maximum
473	274	141	2,885	3,158
16.4	9.3	4.9		

dence of unfavorable ocean conditions with one or all of these trends poses additional and substantial risks to Columbia River salmonid survival or recovery of salmonid populations. Later sections of this report elaborate on the concept of risks and their management in the context of Columbia River flows, withdrawals, and salmon survival rates.

RETURN FLOWS AND WATER QUALITY

In addition to water withdrawals, return flows from irrigation projects like the Columbia Basin Project add to river flows and have implications for Columbia River system water quality and quantity as well as for salmon survival. Complete accounting of surface and subsurface discharges of irrigation return flows from the Columbia Basin Project is not possible because they are not measured. A report from the Montgomery Water Group (1997), however, provides some data from which irrigation return flows can be estimated (the rationale and assumptions made in the mass balance of water in the Columbia Basin Project from 1975 through 1994 are provided in Appendix C). Irrigation return flows from the Columbia Basin Project consist of canal and lateral operational spills, surface irrigation drainage, and groundwater outflow. Canal and lateral operational spills are gauged, but surface irrigation and groundwater outflows to the mainstem are unmeasured and were calculated as the closure (balancing) term in the Columbia Basin Project water balance. From 1975 to 1994, canal and lateral spills averaged 265,000 acre-feet per year and irrigation and groundwater outflow to the river combined to average 540,000 acre-feet per year, for an average total return flow to the Columbia River of 805,000 acre-feet per year. Over this 20-year period of record, 30 percent of the irrigation water for the Columbia Basin Project was thus eventually returned to the river. This also means that 70 percent of the water supply for the Columbia Basin Project was consumed or was evaporated, because the change in water storage in the project was assumed to be zero (see Appendix C).

Several water quality parameters are of key concern in the Columbia River system, including water temperature, dissolved oxygen, nutrients, suspended sediments, pesticides, trace metals, and pharmaceuticals (USGS, 1998). Concerns regarding water temperature are illustrated by summer water temperatures in Crab Creek, a small stream near Beverly, Washington. Crab Creek conveys irrigation return flows from the Columbia Basin Project. Water temperatures in Crab Creek (Figure 3-7) generally reflect variations in air temperature from July to September. Based on daily water temperature records from 1975 through 2002 at Crab Creek, and at the Columbia River at Grand Coulee Dam and near Vernita (Washington Department of Ecology, 2003b), water temperatures in Crab Creek are higher than those in the Columbia River during late winter and spring and lower than Columbia River water temperatures during fall and early winter.

The U.S. Geological Survey's National Water Quality Assessment (NAWQA) Program for the Central Columbia Plateau for 1992 to 1995 reported that "the health of the aquatic ecosystems is substantially affected by agricultural practices and, in a few streams, by wastewater discharges" (USGS, 1998) in Washington and Idaho. Numerous water quality parameters can

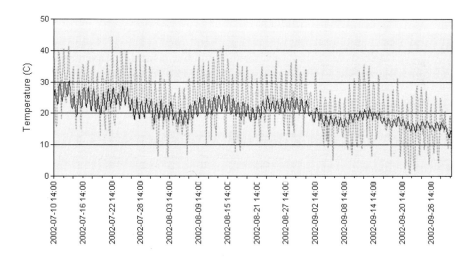

FIGURE 3-7 Water temperatures (bold) in Crab Creek (which conveys return flows from the Columbia Basin Project), and air temperatures, in July, August, and September, 2002. SOURCE: Washington State Dept of Ecology (2003a).

be influenced by agricultural practices and wastewater discharge, including nutrients (specifically nitrogen and phosphorus), sediment, and organic and trace metal contaminants. For example, the NAWQA study found that nitrate-N concentrations increased from less than 1 mg per L in the 1960s to about 3 mg per L in the 1980s (USGS, 1998). Irrigation can also lead to increases in soil erosion and therefore increased sedimentation in streambeds. In the Columbia Basin Project, however, the conversion of surface furrow irrigation to pressurized irrigation (center pivot and sprinkler) since the 1970s (Montgomery Water Group, 1997) has reduced daily suspended sediment yields (load per acre) from about 0.3 pounds per acre in 1975 to about 0.1 pounds per acre in the 1980s (USGS, 1998). In contrast, agricultural return flows in the Yakima River basin have at times contributed to impaired water quality in the Yakima River (U.S. Bureau of Reclamation, 2002). In addition to the influence of sediment on the quality of stream habitat, sediment yield is an important concern because most organochlorine pesticides, such as DDT, are found in streambed sediments. Long banned from use, DDT levels exceeding general standards for aquatic life protection have nonetheless been found in streambed sediments in upper reaches of Crab Creek (USGS, 1998). In the Yakima River basin, DDT and its breakdown products have been found in fish tissue in excess of recommended human health criteria, and concentrations of these pesticides have been correlated with suspended sediment levels (ibid.). Some studies have noted increased cancer risks in certain populations, such as Native Americans, that consume high amounts of Columbia River basin fish (Columbia Basin Bulletin, 2002).

Groundwater samples taken from the Columbia Basin Project exhibit elevated nitrate concentrations, which vary by location. For example, nitrate values in groundwater exceed drinking water standards in shallow groundwater (observation wells), with smaller background concentrations found in deeper wells (USGS, 1998). Pesticide residues were found to be present in high nitrate wells, sometimes exceeding the Maximum Contaminant Level (MCL) for drinking water (USGS, 1998). The U.S. Geological Survey has initiated studies in its NAWQA program to intensively investigate surface and groundwater quality in the Central Columbia Plateau and Yakima River basin (USGS, 2003b).

Beyond immediate human health concerns associated with exposure to pesticides and trace metals through consumption of fish, high concentrations of contaminants and nutrient enrichment in return flows could have long-term implications for the health of salmon populations. For example, in a nationwide reconnaissance of 139 streams conducted in 1999 to 2000 by the U.S. Geological Survey, a wide range of organic wastewater contaminants, including pharmaceuticals and hormones, were detected in streams downstream of sources of human, industrial, and agricultural wastes (Kolpin et al., 2002). Multiple organic wastewater contaminant detection was common, including many compounds for which aquatic life criteria have not been established.

The exposure of fish to organic contaminants, particularly treated municipal sewage discharge, has been demonstrated to impact fish health at several levels, ranging from biochemical processes, to organ functions, to organism fitness (Porter and Janz, 2003). Exposure of adult fish to synthetic hormones and other contaminants with estrogenic properties can significantly impair fertility (Schulz et al., 2003). Exposure to low but detectable levels of organic wastewater contaminants from increasing municipal and agricultural uses of water may thus impact the survival and reproduction of the salmon, especially during summer in low-flow years, when concentrations would be greatest.

WATER TEMPERATURE

There are data for Columbia River water temperatures that date back to 1938 (USACE, 2000. Data provided to Stuart McKenzie for a report sponsored for the Northwest Power Planning Council.). Figure 3-8 shows maximum and average Columbia River August water temperatures at Bonneville Dam. As the figure indicates, the trend lines show clear increases in August water temperatures over time. Water temperature data at Bonneville Dam also reveal that the first and last dates on which water temperatures equal or exceed 20°C are occurring earlier in the year later in the year, respectively. Average and maximum values of Columbia River water temperatures are today both well above 20°C. These increasing trends in water temperatures are

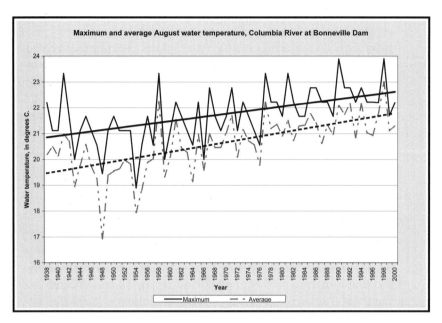

FIGURE 3-8 Maximum and average August water temperatures in the Columbia River at Bonneville Dam (straight lines reflect trends for maximum and average values). SOURCE: USACE, 2000. Data provided to Stuart McKenzie for a report prepared for the Northwest Power Planning Council.

of great concern with regard to the survival of Columbia River salmon. For example, August temperatures at Bonneville Dam exceed temperatures preferred by cold water fish like salmonids (~10° to 15°C; Kling et al., 2003). More importantly, it means that Columbia River water temperatures are approaching the upper limits of thermal tolerance for cold water fishes (~20° to 24°C; Mohseni et al., 2003) such as salmonids.

These temperature changes appear to have been driven by (1) construction of the dam and reservoir system (the large surface areas of Columbia River reservoirs and the increased residence time of water in these reservoirs both contribute to higher water temperatures) and (2) increased temperatures of inflows from tributaries from watersheds that have lost riparian cover that provided shade for those streams.

CLIMATE VARIABILITY AND CHANGE

Changes and variability in regional climate across the Columbia River basin influence discharge and water temperatures of the basin. For example, winter precipitation amounts and snowpack depths in the basin's higher-elevation areas affect seasonal patterns of the river's discharge. Climate variability and changes also have important implications for Columbia River water temperatures (as shown in Figure 3-8). The influences of climatic variability on Columbia River flows have been investigated by many scientists (e.g., Miles et al., 2000; Mote, 2003; Payne et al., 2004), and there is evidence of a gradual warming of Pacific Northwest climate during the twentieth century. For example, in a report for the U.S. Global Change Research Program, a scientific team that evaluated the potential consequence of climate change for the Pacific Northwest concluded that "over the 20th century, annual average temperature in the Northwest rose 1 to 3° F (0.6 to 1.7 ° C) over most of the region" (Mote et al., 1999). The Columbia River was subjected to many changes and human influences during the twentieth century, and care must be taken in ascribing cause-and-effect explanations for climatic and hydrological trends. Some of the concerns regarding possible future climate warming in the region are related to increasing global mean surface air temperature during the twentieth century (about 0.4 ° to 0.8 °C, or 0.7 ° to 1.5 °F; NRC, 2001a). Further evidence of possible regional climate changes might be gained by evaluating climate variability in the undammed Fraser River in Canada, which lies just to the north of the Columbia River basin (see Box 3-2). Climate change and possible warming across the Columbia River basin represent additional uncertainties, such as possible upstream development, tribal water rights adjudications, or variations in ocean conditions, that will affect the cumulative future impacts of water management decisions across the basin.

A key concern regarding possible future climate warming across the basin is the potential effects on the basin's snowpack. Recent research suggests that warmer temperatures across the basin are contributing to declines in total snow accumulations and that the decline in the Cascade Mountains may be as much as 60 percent (Mote, 2003). The implications are that the melting of snowpack earlier in the spring will increase spring runoff

> **BOX 3-2**
> **Canada's Fraser River:**
> **Comparing the Effects of a Changing Climate**
>
> Evaluating the impacts of climate changes and variability is often complicated because the results from other nonclimate variables can intervene and cause similar impacts. For example, water temperatures can be affected by changes in water levels and human activities, such as thermal effects of power plants. In seeking an understanding of how twentieth-century climate affected Columbia River temperatures, a convenient frame of reference exists: Canada's Fraser River. Comparisons between the Fraser and Columbia are useful because the Fraser is relatively close to the Columbia, and its basin is of similar dimensions and has features in common with the Columbia (e.g., headwaters along the western flanks of North America's Rocky Mountains). The Fraser also makes for useful comparison because it is undammed and thus allows for climatic effects on water temperatures in the absence of dams—perhaps the most important human-induced change in Columbia River hydrology—to be considered.
>
> As the figure here illustrates, from 1953 to 1998, the mean summer temperature of Fraser River water temperature increased by 1.1°C (available online at *http://wlapwww.gov.bc.ca/soerpt/*, British Columbia Ministry of Water, Land and Air Protection website). Air temperature in the interior of the Fraser River basin also rose by 1.1 °C in the same period. Fraser River flow has declined since 1913. Although it is not exactly clear what has caused the increases in Fraser River temperature, the increase did not

peaks and reduce summer streamflow.

As Daniel Cayan, a climate scientist at the Scripps Institution of Oceanography explained, "It doesn't mean we've lost water. . . . It means the water is coming off earlier" (quoted in Service, 2004). The upshot is thus that winters would be wetter and summers drier. Not all scientists agree that recent warming across the basin necessarily portends a warmer future, however, as some climate scientists argue that broad trends in temperature and snow accumulation across the basin are due to natural multidecadal oscillations in climate patters.

Many atmospheric scientists are concerned that twentieth-century climate warming in the Columbia River basin was a result of global increases in "greenhouse gases" such as carbon dioxide, and there are some concerns that warming will continue result from dams and their operations. In considering twentieth-century increases in Columbia River water temperatures, data for the Fraser River suggest that the Columbia's temperature increases may not be entirely a result of dams and impoundments and may be affected by other factors such as increasing air temperatures (i.e., climate change).

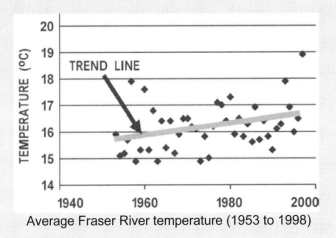

Average Fraser River temperature (1953 to 1998)

SOURCE: Pacific Salmon Commission (1941-1998), Environment Canada, analysis by Canadian Institute for Climate Studies. Figure available online at http://wlapwww.gov.bc.ca/soerpt/pdf/997climate/fraser.pdf.

during the twenty-first century. Atmospheric and climate scientists have developed general circulation models (GCMs) that are used to simulate behavior of the global climate system and to forecast future global and regional changes in climate. Several of these GCMs are used by scientists in North America and Europe, and they are frequently used to forecast regional climatic implications of continued increases in greenhouse gas levels. For example, future Pacific Northwest climate change scenarios from the Intergovernmental Panel on Climate Change and other groups and scientists (Table 3-3) suggest that air temperatures across the region are likely to increase, with less agreement on possible changes in precipitation. In its evaluation of potential climate change impacts on the Pacific Northwest, changes in

TABLE 3-3 Projected Climate Changes over the Columbia River Basin

Area	Winter Temp, °C	Summer Temp, °C	Winter Precipitation	Summer Precipitation, % Change	Year of Projection	Source
Western North America	+4 to +6	+3 to +5.4	0 to +40 % change	-10 to +10% change	2071-2100 minus 1961-1990	Houghton et al. (2001)
Pacific Northwest States	+4.5 to +6	+4 to +4.5	Annual 0 to +50% change		2090	US National Assessment (2000)
Pacific Northwest regional climate	Annual +4.8 to +7.3		-1.0 to +10.4 cm	-4.6 to +2.0 cm	2050	Mote et al. (1999)
Columbia River basin	Annual +1.3		+5 % change		2040-2069	Payne et al. (2004)

precipitation. In its evaluation of potential climate change impacts on the Pacific Northwest, the U.S. Global Change Research Program stated that "regional warming is projected to continue at an increased rate" and also noted less agreement on precipitation forecasts (National Assessment Synthesis Team, 2000). Possible shifts in precipitation patterns and increasing air temperatures have implications for Columbia River hydrology and water management, including water withdrawal permitting decisions. These scenarios represent well-informed speculation on the future, although details across scenarios often show varying results. Nonetheless, the weight of scientific evidence suggests that long-term climate warming of recent decades across the basin is likely to continue. Such long-term temperature increases would represent an increased risk to the survival of Columbia River salmonids, as increasing temperature would represent a threat in terms of further increases in Columbia River water temperatures (which also increased during the latter twentieth century) and reduced flows during low-flow periods. Some observers have noted that the Columbia River water system experiences water system experiences stresses during low-flow periods under current conditions and that "the best water management and planning of today will be done by those with an eye towards both natural patterns of climate variability and possible changes in climate" (Miles et al., 2000). Given the increasing water temperatures in the Columbia River, climate warming across the basin during the late twentieth century, and the prospect of possible additional warming across the basin in the twenty-first century, water management agencies would be well advised to monitor climate data and variability and prepare to adjust operational decisions accordingly as new information becomes available. Appendix D contains additional discussion on climate change and its implications for Columbia River basin hydrology.

SUMMARY

The Columbia River basin experienced a variety of substantial changes to its patterns of water flows and its water quality during the twentieth century. The most dramatic of these changes was the fundamental alteration of the great river's annual hydrograph. At the start of the century, that hydrograph exhibited a great seasonality between its low-flow and high-flow periods. With the construction and operation of the Federal Columbia River Power System, the annual hydrograph was substantially "flattened." By the end of the twentieth century, the differences in flows throughout the year had been greatly reduced, in large part to stabilize flows used to generate hydroelectric power. Considerable interannual and diurnal variability in flows remains, however. Other key operational considerations in the system are flood control, instream flows, and irrigation withdrawals.

There are many users of water along the Columbia River in Washington State. As is the case across the West, irrigated agriculture is the largest consumptive user. About 96 percent of withdrawals are utilized by irrigators, the other 4 percent by municipalities and industries (mainly in the Tri-Cities of Richland, Kennewick, and Pasco). By far the largest irrigator is the federal Columbia Basin Project, which diverts an impressive 74 percent of total irrigation water withdrawals from the Columbia River in Washington. The remaining withdrawals are from a large number of small (relative to the Columbia's flows) withdrawals.

The current pattern of withdrawals is such that they have very little effect on Columbia River flows during January, a period of low demand. By contrast, the volumes of withdrawals in July and August—a period of highest demand—have noticeable effects on Columbia River flows. Although hydrological data on Columbia River withdrawals are imperfect, the data that are available suggest that summer withdrawals in July divert roughly 16.6 percent of river flows at John Day Dam. The upper end of the range of prospective additional withdrawals considered in this study would increase that figure, raising it to roughly 21 percent. A key issue in considering the implications of prospective additional water diversions clearly is the seasonality of those diversions.

Other important changes to the river include deteriorating water quality, which has implications for Columbia River salmon, and increasing water temperatures. Water temperatures in the mainstem Columbia increased steadily during the latter part of the twentieth century. Most observers attribute this increase to the construction of dams and impoundments along the river. Other watersheds in the region that have had fewer hydrological alterations—such as Canada's Fraser River—exhibit increases in water temperature in the absence of impoundments (the magnitude of temperature increases there, however, is smaller than in the Columbia). Prospective climate warming across the Columbia basin may thus also be contributing to this trend. Although precise cause-and-effect mechanisms are hard to define clearly, the changes in Columbia River hydrology identified in this chapter have greatly affected the basin's salmon populations. The following chapter examines relationships between Columbia River salmon and several environmental changes and variables.

4

Environmental Influences on Salmon

Columbia River basin salmon are among the world's most intensively studied fish species. Quantitative and qualitative data regarding salmon species and their habitat have been gathered and evaluated for many decades. This information has increased understanding of Pacific salmon and their complex life histories. Given their responsibilities to help protect salmon, water management agencies in the Pacific Northwest have drawn heavily on this information and have consulted with fisheries scientists in designing strategies for preserving and enhancing salmon habitat and populations. Despite the extent of data and scientific knowledge regarding Pacific salmon, more precise understanding of salmon is inhibited by the complexities of salmon's diverse anadromous (which refers to organisms that spend most of their adult lives in saltwater and then migrate to fresh water and lake to reproduce) life histories and the vast scale of the biomes they traverse during their life spans.

In addition to the biological complexities of salmon species, within the impounded Columbia River they have been affected by an array of environmental conditions and changes, such as increasing water temperatures and changes to other water quality parameters, changes to water velocity through reservoirs, habitat degradation, changing turbidity, shifting seasonal patterns and volumes of river flows, passage effects at dams, and changes in predators and predation rates. Scientists and water managers have considered these issues when formulating fish passage strategies such as flow augmentation, construction of smolt (young salmon, generally two to three years in age) bypass systems, spill programs, smolt transportation programs, and the construction and upgrade of fish ladders. Collectively, these devices and strategies are designed to work in concert to increase survival rates of salmon migrating through the dammed river and contribute to the productivity of anadromous fish populations. NOAA (National Oceanographic and Atmospheric Administration) Fisheries (formerly the National Marine Fisheries Service, or NMFS), the federal fishery agency responsible for the recovery of anadromous salmonid populations listed pursuant to the Endangered Species Act, embraces these strategies and calls for their continued improvement and use in fostering salmon recovery (NMFS, 2000). Even so, it is not known whether these actions alone can reverse or stall long-term declines in salmon populations. Much of the research identified in the 2000 Biological Opinion from the NMFS focuses on improving the implementation of these strategies and gaining a clearer understanding of the outcomes of management actions that are often confounded by environmental complexities. Furthermore, conditions in tributaries and in estuarine and marine habitats have pronounced effects on salmon productivity, as do harvest and hatchery programs. Large salmon returns in 2001 to 2003, for example, were viewed by many scientists as a function of favorable ocean conditions (NPCC, 2003), but ecological and biological complexities inhibit perfect understanding of cause and effect in such events. In any event, a 100-year snapshot of Columbia River salmon portrays long-term declines and provides a backdrop against which short-term events should be evaluated. This chapter reviews environmental variables that affect Columbia River salmon and examines competing hypotheses and models constructed to explain the relative importance of these variables.

COLUMBIA RIVER SALMON

Three species of anadromous salmonids commonly migrate through the middle and upper reaches (above Bonneville Dam) of the Columbia and Snake rivers in the State of Washington: Chinook (*Oncorhynchus tshawytscha*), steelhead (*Oncorhynchus mykiss*), and sockeye (*Oncorhynchus nerka*) all commonly migrate to spawning destinations well upstream from Bonneville Dam. Remnant wild and hatchery populations of coho salmon (*O. kisutch*) are also found in select locales in the upper Columbia basin. All these species have some population units that are listed as endangered or threatened under the Endangered Species Act (see Table 1-1). Additionally, chum salmon (*O. keta*), which are also federally listed, and a vestigial population of pink salmon (*O. gorbuscha*), inhabit waters downstream from Bonneville Dam.

Requirements for each stage of salmon life history can be generalized for all of the anadromous species. Spawning fish, returning from the ocean, require freshwater instream habitat with temperatures that ensure survival until they spawn. Spawning salmon seek species-specific gravels, water depths, and velocities to build redds (nests) in which they deposit their eggs. Egg survival depends on low sedimentation rates, adequate delivery of dissolved oxygen, and appropriate river temperatures to support egg development. Once the eggs hatch, some of the young fish (fry) maintain locations in the river to develop, while some fry grow while migrating downstream. During the post-fry stage (juvenile), these fish remain in the river from several months to more than two years, depending on the species or life history type. Growth is crucial during this phase, which supports the physiological transformation required for emigrating from fresh water, into brackish water, and then into saltwater. This transformation phase is called *smoltification* and during it the fish undergo a complex physiological process that prepares them for adaptation to seawater as they migrate downstream (as their names suggest, spring migrants experience smoltification during spring months, and summer migrating ocean-type Chinook go through smoltification mainly in July and August).

Chinook Salmon

Fishery managers traditionally divide Columbia River Chinook salmon into spring, summer, and fall runs. After spending much of their lives in the Pacific Ocean, spring Chinook salmon adults that spawned in high, cold tributaries in Idaho, Oregon, and Washington return to the Columbia River mouth from February through mid-May. Through olfactory homing instincts, they travel upstream and reach their natal tributary streams in June, move to spawning sites in August, and largely complete spawning by early September. Summer Chinook salmon, which use the Columbia River upstream from the mouth of the Snake River, enter the river mostly in May and June and spawn in September and early October in natal streams such as the Wenatchee and Methow rivers. In the Snake River, summer Chinook salmon make up a later component of the spring Chinook salmon migration, spawning in late August and early September. Fall Chinook salmon enter the Columbia River in July and August and spawn in late October and November in the mainstem river (a small number also spawn in the Snake River between Lewiston and Hells Canyon Dam). Fall Chinook salmon today make up the largest segment of Chinook salmon runs.

Hatchery and naturally produced fall Chinook salmon that use the lower Columbia River area are known as "tule" fall Chinook salmon. Relatively dark in color, they arrive in the river in September and October, then spawn in late fall. Fall Chinook salmon that spawn upstream from McNary Dam in both the Snake and Columbia rivers are known as "upriver brights."[1] They enter the Columbia River in August and spawn mostly upstream from McNary Dam. Upstream from Bonneville Dam, the (numerically) most important spawning area—a long, damless stretch of river known as "The Hanford Reach"—lies between Priest Rapids Dam and the head of McNary Dam pool.

The shoreline-oriented behavior of subyearling fall Chinook salmon in flowing river segments, and their relatively slow rearing migration in the Snake and Columbia rivers, which occurs in early and midsummer, makes them potentially vulnerable to high water temperatures. Construction of mainstem hydroelectric projects, and the consequent slower river velocities, extended the passage period for subyearling (juvenile fish less than one year old) fall Chinook in the Hanford Reach (Chapman et al., 1994; Park, 1969). Reservoirs like McNary and Lower Granite pools, however, may serve as surrogates for estuarine rearing (Chapman et al., 1994). Fall run Chinook usually migrate to the ocean during their first spring and summer in fresh water. Most yearling spring Chinook salmon migrate in April and May and reach the estuary in early June of their second year in fresh water, thus evading the warmest Columbia River waters of early and midsummer. Fall run and spring run Chinook are often called ocean and streamtypes, respectively. Returns of spring Chinook and Snake River "summer" Chinook are dominated by hatchery-reared fish. Returns of fall Chinooks (upriver brights) are pri-

---

[1] "Brights" also describes fall Chinook that spawn in the Lewis River, a Cowlitz River tributary, and in the Deschutes River.

marily wild fish.

Steelhead

Columbia River steelhead are categorized according to two broad modes of behavior. *Winter steelhead* remain at sea until late fall or winter, then enter the Columbia River and tributaries as far upstream as Fifteen Mile Creek at The Dalles, which enters the Bonneville Dam pool. They spawn in late winter and early spring, and fry emerge from redds in late spring to July. Juveniles spend two winters in fresh water before migrating to sea in March to early June. *Summer steelhead*, by contrast, which use some tributaries downstream from Bonneville Dam (e.g., Kalama River) and virtually all suitable streams upstream from Bonneville, enter the Columbia River from May to early September. Adults spend the winter in the mainstem of the Columbia and Snake rivers and in large tributaries and spawn mostly in the period from March to May. Like winter steelhead, fry emerge from redds in late spring to midsummer and spend at least two winters in fresh water before migrating to sea as smolts. The smolts move seaward in spring. Returns of steelhead at the Columbia River estuary are dominated by hatchery-reared fish.

Sockeye Salmon

Sockeye salmon require a lake for juvenile rearing. Sockeye were once found in the upper Columbia River lake and tributary systems of the upper Columbia River upstream from Grand Coulee, in Suttle and Wallowa lakes in Oregon, in the chain of Okanogan River lakes and Lake Wenatchee, and in the Stanley basin lakes of the upper Salmon River in Idaho. They spawn in fall upstream from the two lakes, and fry move downstream soon after emergence from redds, rearing in the lake environment for mostly one but sometimes two years. As smolts they emigrate in April and May. Sockeye currently inhabit only Osoyoos Lake in Canada, Lake Wenatchee in Washington, and Redfish Lake in Idaho. Sockeye salmon return to the Columbia River estuary mostly in May and June. The bulk of these returns are wild fish.

Coho Salmon

Coho salmon in the Columbia River mostly spawn (and juveniles rear) in tributaries downstream from The Dalles Dam. Hatchery-produced coho predominate. Wild coho formerly used a number of other tributaries, including some upstream from McNary Dam, like the Yakima, Methow, and Grande Ronde rivers. Most coho smolts move seaward in the spring.

Variations in Migratory Patterns

These different salmon and steelhead species and subspecies migrate downstream and upstream through the Columbia River system at different times of year. The greatest risks to the survival of migrating fish occur during periods when Columbia River temperatures are highest and during low-flow periods and in low-flow years. Species and life stages of listed fish that transit the Columbia River mainstem in summer months (June to August) include:

1. Subyearling fall Chinook from the Snake River;
2. Late-migrating steelhead (smolts);
3. Snake River adult sockeye salmon (adults);
4. Snake River summer Chinook (adults);
5. Snake and Columbia river steelhead (adults);
6. Snake River fall Chinook (adults); and
7. Bull trout.

This report contains several references to the risks of survival of Columbia River salmonid stocks during critical periods. References to fish in the system during these periods do not apply to all salmon and steelhead species and subspecies but rather focus on the species listed here that transit the system during the critical June-August period.

STATUS OF SALMON AND STEELHEAD STOCKS

Historical perspectives of trends in Columbia River salmon abundance are essential to understanding the relative abundance of recent and current salmon runs as well as long-term fishery trends. Many sources of data contribute to scientific knowledge of historical changes in the abundance of the Columbia's anadromous salmon and steelhead. Because of their abundance (and their size) in the Columbia River, Chinook salmon have long attracted the attention of fishery scientists and have been intensively monitored and tracked over time. Fish counts at Bonneville, McNary, Priest Rapids, and Lower Granite dams for the period 1977 to 2002 (Figures 4-1, 4-2, and 4-3, for adult Chinook, adult steelhead, and adult sockeye, respectively) provide an overall picture of changes in the status of salmon populations over time.

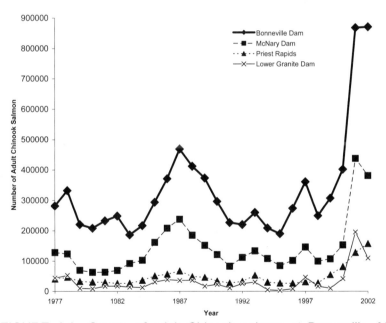

FIGURE 4-1 Counts of adult Chinook salmon at Bonneville, McNary, Priest Rapids, and Lower Granite dams on the Columbia River (1977 to 2002). SOURCE: Fish Passage Center (available online at *http://www.fpc.org/adult_history/adultsites.html,* last accessed November 17, 2003).

Returns of Chinook from 2001 to 2003 greatly exceeded the 1993 to 2002 average returns (Figure 4-1) and generated a great deal of excitement in the Pacific Northwest. These record returns have generally been attributed to favorable ocean conditions. The Northwest Power and Conservation Council, for instance, asserted that "good ocean conditions are creating strong adult returns" and noted that "ocean conditions will change" (available online at *http://nwppc.org/news/2003_11/3.pdf*, last accessed December 2, 2003). It bears noting that the 2001 to 2003 returns of fall Chinook salmon, like in-river runs since the mid-1990s, also benefited from increased restrictions on ocean fishing. In addition to recent, comparatively large Chinook runs, steelhead returns also rose sharply relative to figures since the mid-1970s (Figure 4-2). Sockeye also experienced an increase in returns in the late 1990s (Figure 4-3).

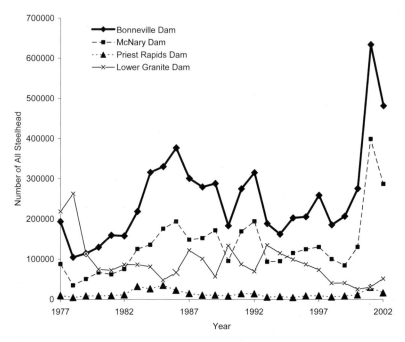

FIGURE 4-2 Counts of all adult steelhead at Bonneville, McNary, Priest Rapids, and Lower Granite dams on the Columbia River (1977 to 2002). SOURCE: Fish Passage Center (available online at *http://www.fpc.org/adult_history/adultsites.html,* last accessed November 17, 2003).

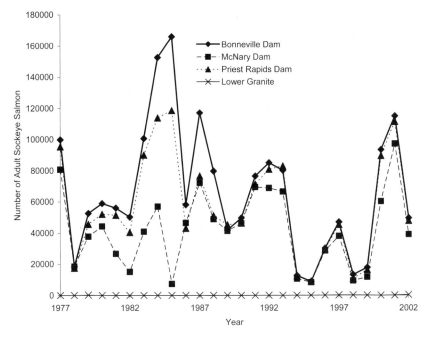

FIGURE 4-3 Counts of adult sockeye salmon at Bonneville, McNary, Priest Rapids, and Lower Granite dams on the Columbia River (1977 to 2002). SOURCE: Fish Passage Center (available online at *http://www.fpc.org/adult_history/adultsites.html,* last accessed November 17, 2003).

Redd counts from Idaho's Salmon River basin provide additional information regarding temporal trends of spring/summer Chinook salmon listed by the Endangered Species Act.[2] Redd counts in 1957, the first year of systematic surveys, were inflated by completion of The Dalles Dam in the lower Columbia River (Figure 4-4). The reservoir behind the dam flooded the Celilo Falls, which was an important Indian fishing site. As a result of the loss of this important fishing site and an attendant reduction of harvests, Columbia and Snake river escapements of salmon and steelhead increased sharply. Later, as Indian fishers shifted to gillnets, fishing and harvest rates increased.

---

[2] "Summer Chinook" salmon in Idaho, like spring Chinook salmon, spend one winter in natal tributaries before migrating to sea. They spawn principally in the South Fork Salmon River and upper Salmon River.

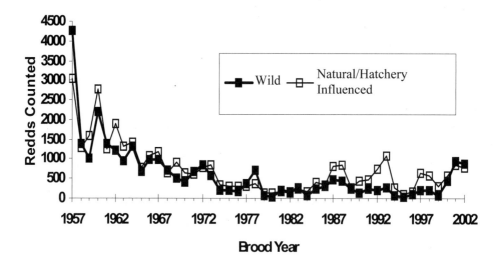

FIGURE 4-4 Number of combined spring and summer Chinook redds (thousands) counted in Salmon River drainage, wild and natural/hatchery-influenced trend areas, 1957-2002. SOURCE: Fish Passage Center (available online at http://www.fpc.org/adult_history/adult-sites.html, last accessed March 24, 2004).

Figures 4-5 and 4-6 present a longer time frame of reference of salmon abundance and its changes, and they reflect a steady decline in the spring Chinook catch since the early 1940s (there are, however, some departures from this long-term trend, such as increases in landings in the mid-1980s). The harvest rate in the Columbia River between the river mouth and the upper limit of commercial fishing near the site of McNary Dam ranged from 40 to 85 percent before the 1960s, declined until 1974, and thereafter averaged less than 10 percent (Chapman et al., 1995). Numerical harvest in the post-Bonneville Dam era peaked in the 1950s, declined to 1974, and then remained negligible. Declines in salmonid stocks, and the variations in declines across stocks, have been described as follows:

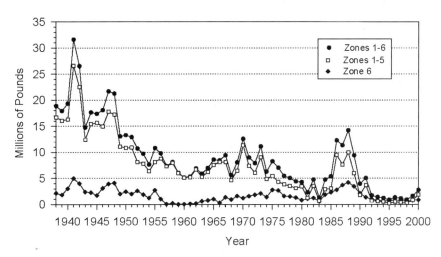

FIGURE 4-5 Commercial landings of salmon and steelhead from the Columbia River in pounds, 1938 to 2000. SOURCE: WDFW-ODFW (2002).

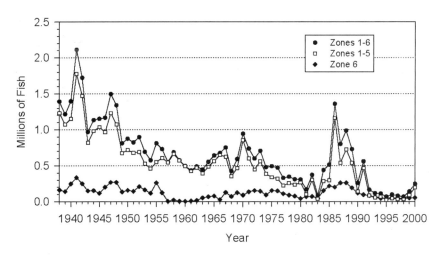

FIGURE 4-6 Commercial landings of salmon and steelhead from the Columbia River in numbers of fish, 1938 to 2000. SOURCE: WDFW-ODFW (2002).

> The Columbia has numerous kinds and runs of salmon and not all runs have declined at the same pace. There are yearly variations. There are temporary recoveries for some species and runs, but overall the decline has been pervasive and general. The catches on the Columbia are one measure of the decline. From 1880 to 1930 the catch was 33.9 million pounds a year. From 1931 to 1948 it declined to 23.8 million. From 1949 to 1973 the yearly average fell to 10.9 million pounds. In 1993 the catch was 1.4 million pounds. (White, 1995, p. 97)

Populations of the basin's anadromous fish stocks are currently estimated to be generally less than 10 percent of their typical historical levels (Chapman, 1986; Kaczynski and Palmisano, 1993; NPPC, 1986).

In addition to historic declines, another important change is an increasing proportion of hatchery-reared fish in the salmon population. The majority of spring Chinook salmon, summer Chinook salmon, and steelhead counts in recent years showed that most of these fish originated from hatcheries. Only about one-fourth or less of spring/summer Chinook salmon and steelhead that returned to the Snake and upper Columbia rivers in the past two decades have been of wild origin; thus, about 75 percent of the spring/summer adult Chinook salmon that return to the Snake River are produced in hatcheries. The proportion of wild fish in the salmon population is an issue important to long-term survival of the species, as pointed out by a previous National Research Council committee that reviewed Columbia River salmon populations and management: "The long-term survival of salmon depends crucially on a diverse and rich store of genetic variation. . . . Management must recognize and protect the genetic diversity within each salmon species. . . . It is not enough to focus only on the abundance of salmon" (NRC, 1996).

In summary, salmon populations of the Columbia River have decreased dramatically since the 1800s, albeit with annual variations in abundance. Although returns of Chinook salmon and steelhead increased sharply from 2001 to 2003 relative to the 1975 to 2000 numbers, they remained but a small fraction of former abundance. Furthermore, fish of hatchery origin from a few stocks constituted most of the runs of spring and summer Chinook salmon and summer steelhead. Genetic diversity within these salmon runs has thus declined, which may have reduced the potential for these species to adapt to environmental changes, such as warmer water temperatures (Brannon et al., 2002).

RESEARCH, MODELING, AND ALTERNATIVE HYPOTHESES

Flow Augmentation

The Federal Columbia River Power System consists of a vast network of storage reservoirs and run-of-river dams, connected in some areas by undammed river segments. Prior to 1983, water in the system was primarily managed to accommodate and balance a variety of demands that included flood control, hydropower, recreation, irrigation, and other extractive demands. In 1983, as part of the Northwest Power Planning Council's Fish and Wildlife Program, a flow augmentation strategy was developed. The program provided for an allotment of water directed specifically at increasing instream flows during the period that smolts migrate seaward. The amount and timing of these releases, known as the Water Budget, was determined annually. The Water Budget has subsequently evolved into a more extensive and complex water management strategy intended to increase instream water velocities, reduce travel times, and increase survival rates of smolts as they migrate seaward through the impounded Columbia and Snake rivers (spring migrants smolt during the spring months, and summer-migrating ocean-type Chinook migrate primarily in July and August). This water management strategy is referred to as flow augmentation (NMFS, 2000). Releases today are made after considering requests from the Fish Passage Center in Portland, which represents fisheries agencies and tribal groups. Implementation of this strategy has reshaped the pre-1983 annual hydrograph, resulting in more pronounced peaks during the spring and summer smolt migration periods. The demand for instream flows is an important priority and is a prominent action and feature in the 2000 Biological Opinion of the NMFS. Not surprisingly, this new demand has impacted other water management needs throughout the system, and has necessitated a new balance among system users.

Rationale for Flow Augmentation

Flow augmentation is the directed release of water from storage reservoirs to increase instream flows, which are intended to help reestablish suitable migratory conditions for smolts that migrate seaward through the impounded Snake and Columbia rivers; flow augmentation from Dworshak Reservoir on the Clearwater River in Idaho is also used to add cold water to the Lower Snake River. Flow augmentation from the Columbia River is provided from several large storage reservoirs. These include Grand Coulee reservoir (Franklin D. Roosevelt Lake) and a complex of storage reservoirs in Canada and Montana. In the Snake River basin, Dworshak reservoir, Brownlee reservoir, and the Hells Canyon complex—all in Idaho—augment flows (Figure 4-7). The rationale for flow augmentation is founded on two premises:

1. Increased discharge results in higher water velocity through reservoirs which, in turn, increases the migration speed of smolts in the impoundments of the Lower Snake and Columbia rivers, ultimately resulting in increased smolt survival through this migratory corridor.

2. Increased discharge lowers water temperature, improving migratory and rearing conditions for both juvenile and adult salmonids, particularly during the summer.

Smolt Survival

Cada et al. (1997) reviewed literature from within and outside the Columbia River basin, addressing the influence of water velocity on the survival rates of juvenile salmon and steelhead. Most of the studies reviewed identified a positive relationship between outmigration flows and survival but noted substantial uncertainty regarding many of the estimates. In many cases the relationships described did not consider interactions with factors other than water velocity. Other factors examined in the review included predation, water quality, and physiological state of the smolts at the time of migration. Despite limited data, Cada et al. found that a general relationship of increasing smolt survival with increasing flow in the Columbia River basin was a reasonable conclusion.

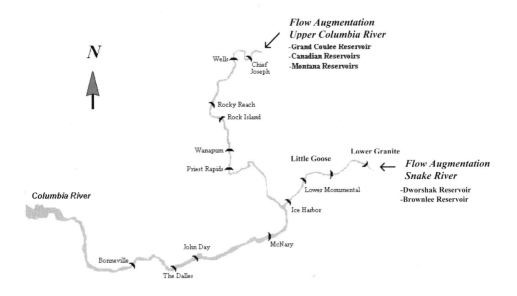

FIGURE 4-7 Dams and impoundments on the Snake and Columbia rivers, through the reaches where anadromous fish passage is accommodated. Sources of flow augmentation water are indicated. SOURCE: Reprinted from Giorgi et al. (2002).

The migration speed of salmon smolts dictates their exposure time to hazards in reservoirs. For example, predatory fish and birds are responsible for a substantial amount of smolt mortality incurred within the impounded Columbia River. Northern pikeminnow, smallmouth bass, channel catfish, and walleye all prey heavily on smolts. It has been estimated that the predacious northern pike minnow consumed 78 percent of the smolts that were lost to predatory fish in John Day reservoir from 1983 to 1986 (Rieman et al., 1991). In the 1990s a control program (in the form of a bounty fishery) that targets these species was implemented (Young, 1997a and 1997b). Birds also consume large numbers of smolts at various locations throughout the Columbia River. An expanding Caspian tern population and double-crested cormorants are effective smolt predators in some areas downstream of Bonneville Dam. Gulls also prey upon smolts in the tailraces (outflows below dams) of Columbia River dams

(Collis et al., 2002). Rugerrone (1986) estimated that in 1982, gulls foraging in the tailrace of Wanapum Dam consumed 2 percent of the smolts passing the dam. In an effort to reduce smolt mortality, a variety of actions have been directed at displacing, harassing, or excluding predatory birds from problem areas.

Historical Background

Shortly after the construction of several Snake River dams, federal biologists documented that dams and associated reservoirs delayed the migration of smolts. For example, Ebel and Raymond (1976) and Bentley and Raymond (1976) estimated that after dam emplacement, travel times of yearling Chinook salmon and steelhead increased at least twofold over preimpoundment conditions. The first explicit depiction of a flow-smolt survival relationship was presented by Sims and Ossiander (1981). Building on previous studies (e.g., Raymond, 1979; Sims et al., 1976, 1977, 1978), Sims and Ossiander (1981) constructed a series of graphs depicting that annual indices of smolt migration speed and survival were positively correlated with annual indices of flow and spill volumes during migratory periods (1973 to 1979). Although it was not possible to separate reservoir effects (associated with migration speed) from passage effects attending spill passage, this was the first evidence establishing the flow-travel time-survival relationship. Furthermore, these findings were the foundation that led to the development of both the flow augmentation and spill programs in place today. Both spill and migration speed were defined as agents affecting smolt survival. Shortly thereafter (in 1983), the "Water Budget" was established by the Northwest Power Planning Council. Under that program, a specific volume of water in Snake River storage reservoirs was dedicated to flush smolts seaward. The Fish Passage Center (previously known as the Water Budget Center) in Portland provides fish passage technical advice regarding spill, flow, and fish facilities operations to fish and wildlife managers was established to track the delivery of water and the response of smolts to the water management action (see *http://www.fpc.org/*, last accessed March 13, 2004). That original water management strategy expanded over the ensuing two decades to the current flow augmentation program described in the 2000 Biological Opinion from the NMFS.

Throughout the 1980s smolt travel time was consistently monitored. In the early 1990s, studies concluded that variability in smolt travel times was best explained as a function of a combination of flows, water temperatures, and release dates (the latter of which is a surrogate for the level of smolt physiological development; Berggren and Filardo, 1993). It was reported, however, that average river flow explained most of the observed variability in smolt travel time for most stocks investigated (ibid.). These findings reinforced the strategy to provide flushing flows to increase migration rates.

During the same period, federal scientists investigated the migration of ocean-type subyearling Chinook salmon through the John Day Pool (Giorgi et al., 1994). Their characterization of migratory behavior in John Day Reservoir differed from that described by Berggren and Filardo (1993). They did not identify a consistent relationship between smolt travel time and any of the three predictor variables (flow, water temperature, or release date), but rather characterized the migratory patterns as a complicated mix of rearing and migratory behavior, often punctuated by extensive upstream excursions.

Williams and Matthews (1995) questioned the foundation of the Sims and Ossiander (1981) flow-survival relationships by asserting that the 1970s-era data reflected operating conditions that no longer existed in the contemporary hydrosystem. They suggested that the high smolt mortality rates witnessed during low-flow years in that era was in part associated with slow rates of migration through the system, but was exacerbated by turbine and powerhouse operations. Furthermore, they concluded that the Sims and Ossiander flow-survival relationship does not accurately predict the survival of spring-migrating smolts under contemporary hydrosystem operations and the smolt bypass systems in place at dams. The research community generally recognized the need for statistically robust survival estimates acquired in the contemporary setting, since the flow-survival debate was intensifying as more water was being shifted toward flow augmentation. But sampling limitations associated with the need to handle and inspect large numbers of freeze-branded smolts prevented the use of new analytical methods reported by Burnham et al. (1987).

Over recent decades, technological improvements have allowed for more accurate smolt survival estimates. The advent of the passive integrated transponder (PIT) tag, and associated detection systems that could be retrofitted to existing smolt bypass systems, fostered the transition to a new era and quality of smolt survival and travel time estimates for the Columbia-Snake river system (Prentice et al., 1990). Since 1994, smolt survival estimates have been obtained through segments of the Federal Columbia River Hydro System by the NMFS/NOAA Fisheries. The bulk of the data for use in flow-survival assessments are from the Lower Snake and, to a lesser extent, portions of the lower Columbia. There is a paucity of data available for the middle reach of the Columbia River upstream from McNary Dam. Even now, with widespread use of PIT tags, opportunities to provide robust smolt survival estimates through the middle reach of the Columbia River are limited because of the small number of PIT detection systems there.

Contemporary Investigations

Translating river flows, or smolt migration rates, into smolt survival rates is the critical issue underpinning the rationale for providing flow augmentation and quantifying any associated benefits. A great deal of research since 1994 has been directed toward a better understanding of these complex relationships. During the 1990s, research increasingly focused on identifying a more complex suite of factors that influenced migration speed through the hydrological system. The collective research indicated that the species responded differently to various factors through different segments of the river. In both the Snake and Columbia rivers, yearling Chinook salmon migration speed was correlated with both flow (water velocity) and the level of smolt development (Beeman et al., 1991; Giorgi et al., 1997; Muir et al., 1994). River discharge (flow) was determined to be the factor that explained the majority of variability in migration speed for steelhead (Buettner and Brimmer, 2000; Giorgi et al., 1997) and sockeye salmon (Giorgi et al., 1997).

The modern era of smolt survival studies continued in the Snake River and in portions of the lower Columbia River, since an extensive network of PIT detections systems is located there (most flow-survival studies have been conducted in the Snake

River, and results from the Snake are generally felt to reflect processes that occur elsewhere in the system). Scientists from NOAA Fisheries generally design and conduct those studies, but the agency relies on the broad-based PIT-tagging program overseen by the Columbia Basin Fish and Wildlife Authority (CBFWA, a coalition of tribes, and state and federal wildlife management agencies) to provide tagged fish for monitoring. Smith et al. (2002) used multiple regression methods to assess the effects of a variety of factors on smolt migration rate and survival for 1995-1999. Using a mixture of PIT-tagged yearling Chinook salmon and steelhead smolts from the Snake Basin, they found that travel time from Lower Granite Dam to McNary Dam was strongly correlated with flow volume, with the physiological development of the smolts a contributing factor, particularly for Chinook salmon. However, they could not identify a substantive or consistent relationship between smolt travel time and smolt survival through that same river segment. The authors concluded that survival benefits from increased flow were minimal at best, and that any benefits may be expressed downstream from McNary Dam, beyond their observation zone. These findings were consistent with those expressed in an earlier "White Paper" (NMFS, 2000), which assessed flow, migration speed, and smolt survival.

Drought conditions in 2001 created one of the lowest runoff years on record for the Columbia River, which presented an opportunity to monitor smolt survival under low-flow conditions. Consistent with the findings of Smith et al. (2002), Zabel et al. (2002) found no flow-survival relationship for yearling spring and summer Chinook salmon (1993 to 2001). The Zabel et al. group found that smolt travel time was correlated to river discharge volume, but no relationship between migration speed and survival was evident. Survival was depressed in 2001 relative to many other recent years; however, low flows were not the only factor implicated in low survival rates through the hydrosystem, as spill was minimal or nonexistent at most dams that smolts encountered. Both conditions likely contributed to poor survival. Furthermore, water temperature has been implicated as a principal factor affecting smolt survival, particularly in low-flow water years, when seasonal water temperature increases earlier and to higher levels (Anderson, 2003).

Zabel et al. (2002) suggested that even in the absence of a

flow or migration rate-survival relationship, other benefits may be provided by the swifter migration made possible by increased flow levels. They speculated that higher flows may improve estuary and Columbia River plume conditions and associated survival through those zones but offered no empirical evidence for such. In contrast to yearling Chinook salmon, steelhead survival figures dramatically decreased in 2001 compared to figures for the 1990s. Three factors were implicated as causing this dramatic increase in mortality of Snake River steelhead. First, spill was negligible at most of the dams the steelhead encountered. This mechanism is distinct from migration speed-related processes. Second, of all the salmon species, steelhead migration speed appears to be the most sensitive to flow and associated water velocity (Berggren and Filardo, 1993; Giorgi et al., 1997). Lastly, water temperatures warmed sooner in 2001 than in the preceding three years (see *http://www.cbr.washington.edu/dart/dart.html*, last accessed February 28, 2004). This pattern was evident in both the lower Snake and Columbia rivers. Water temperatures exceeded 12.5°C early (by the first week in May at Lower Monumental Dam) in the steelhead migration. This, coupled with slow migration speed, can compromise steelhead migratory processes. Increasing water temperature can disrupt the migratory behavior of steelhead and foster reversion to the fresh water parr (a young salmon during its first two years of life, when it lives in fresh water) state. It is plausible that if migration rates are slowed (as witnessed in 2001; see Zabel et al., 2002), steelhead smolts may have been exposed to seasonally increasing water temperatures that exceeded the threshold to support smoltification and thus they remained in the mainstem.

The Fish Passage Center also monitors smolt migration throughout the system and provides estimates of smolt survival through the hydrosystem. The center's characterization of flow–survival dynamics differs from that of investigators from NOAA Fisheries. The center expressed its conclusions in a paper submitted to the (previous) Northwest Power Planning Council (FPC, 2002), stating that for juvenile steelhead and Chinook salmon spring migrants

- a water travel time and survival relationship exists for spring migrating Chinook salmon and steelhead of Snake River and mid-Columbia River origin;

- a water travel time and fish travel time relationship exists for spring migrating Chinook salmon and steelhead; and
- it is difficult to define a flow-survival relationship because survival is the combined result of many interacting variables and the method for estimating survival does not lend itself to identifying each environmental or biotic variable individually.

Snake River Fall Chinook Salmon

For fall Chinook salmon in the Snake River, flow, water temperature, and turbidity are correlated with migration speed and survival (Smith et al., 2003). Over the course of summer migration, river discharges decrease, temperatures increase, and turbidities decrease. Thus, predictor variables were typically correlated among themselves. In the middle reach of the Columbia River, the size of subyearling Chinook salmon was found to be the best predictor of migration speed between Rock Island and McNary dams (Giorgi et al., 1997).

John Day Project (McNary tailrace to John Day tailrace)

Smith et al. (2002) also examined survival dynamics of fall Chinook salmon from the tailrace of McNary Dam to the tailrace of John Day Dam. Fall Chinook salmon were collected, PIT tagged, and released at McNary Dam. The population was primarily composed of mid-Columbia River stocks, such as the wild population from Hanford reach. They found that during the summer (1998 to 2001) correlations were not significant between annual survival and the average river condition variables measured at McNary Dam, but the correlation with temperature was considerably higher than for flow and turbidity.

Northwest Power and Conservation Council's Independent Science Advisory Board Studies

In an effort to shed some light on this complex and often contradictory mass of information, the (previous) Northwest Power Planning Council called on its Independent Science Advisory Board (ISAB) to review, update, and clarify the effectiveness of flow augmentation. The ISAB challenged the results from the prevailing flow/smolt survival model that spurred post-1983 formulation of smolt-migration water policy, concluding that "the prevailing flow-augmentation paradigm, which asserts that in-river survival will be proportionately enhanced by any amount of added water, is no longer supportable. It does not agree with information now available" (ISAB, 2003). Support for this recent conclusion was based largely on datasets acquired in the lower Snake River from the Lower Granite Project to McNary Dam on the Columbia River. They relied heavily on survival estimates and analyses from NOAA Fisheries to characterize the spring period and information from the U.S. Fish and Wildlife Service and from the Fish Passage Center to describe a survival model for the summer period (the models described in this section are primarily based on regression analyses. Also see *http://www.nwcouncil.org/ fw/science.htm*, last accessed March 15, 2004, for more information on ISAB models and studies.)

Flow-survival. The ISAB presented a "broken-stick" flow-survival model to describe the NMFS-generated PIT survival data it reviewed (ISAB, 2003). That is, the board identified a "breakpoint" near 100,000 cfs for yearling Chinook and steelhead in the Snake River during the spring. According to this report, when flows exceed that threshold, no flow-survival relationship is apparent. The value of flow augmentation is thus questionable above those levels. Below that breakpoint, a flow-survival relationship is evident. However, the report did not derive algorithms to describe the two legs of the generalized model but rather depicted the model graphically. The intent is apparently not to offer this as a predictive tool but rather as a visual framework to introduce the new hypothesis.

Survival dynamics below the breakpoints. With respect to the lower survival rates observed below the breakpoints presented in the 2003 ISAB report, the board hypothesized that specific hydropower operations in the form of daily load-following cycles create hydraulic dynamics that affect survival, rather than average daily flow discharged through the complex of reservoirs and dams ("load following" refers to adjustments in power production to meet changes in power demand or "loads"). They noted that the frequency and intensity of load following substantially increase when river discharge falls below the breakpoints. They suggest that diminishing or eliminating load following will improve smolt survival more than merely providing higher average daily flows. According to the board's hypothesis, the hydrological effects of load-following power generation disrupt migration cues, which ultimately results in lower smolt survival during migration.

Fall Chinook salmon summer model. The emphasis in this model is also on the Snake River. In formulating the summer model, weekly survival estimates for ocean-type subyearling Chinook migrating from release sites upstream to the tailrace of Lower Granite Dam, as estimated by the Fish Passage Center for the years 1999 and 2000, were employed. As was the case for the spring model, the summer model is described only in generic terms, with "breakpoints" between two legs near 40,000 and 50,000 cfs. The ISAB report offered new hypotheses for describing smolt survival patterns observed in the Snake River. But it cannot be certain that a "broken-stick" model accurately explains survival patterns in the mainstem Columbia River, as no direct evidence to support such in that river segment was provided. Analyses of flow-travel time relationships have been published and cited by the ISAB for the middle reach of the Columbia River, but no definitive flow-survival analyses were ever published. The paucity of robust, consistent survival indices for the Columbia River thus limits meaningful survival analyses with respect to prevailing environmental conditions.

The ISAB report received immediate attention. The CBFWA staff drafted a 34-page technical memorandum commenting on the ISAB assertions and hypotheses (CBFWA, February 26, 2003), which contained a cover letter stating:

> In conclusion, we believe that the ISAB report supports the biological rationale for the minimum flow objectives contained in the NMFS Biological Opinion. The ISAB report presents additional hypotheses for future study that are of some interest, although there is little data at the present time to support these hypotheses. The ISAB does suggest some operational changes in river operation that may offer benefits when Biological Opinion flow objectives cannot be met, which warrant further study and consideration.

The CBFWA group challenged, however, the ISAB characterization of the flow augmentation, noting:

> We do not agree with the ISAB's characterization of the flow augmentation paradigm, which they state, "asserts that in-river smolt survival will be proportionately enhanced by any amount of added water." Establishing reservoir draft limits and augmenting base flows with additional water are only the tools whereby the objective of providing migration flows is accomplished.

The CBFWA group questioned whether altering load-following operations can adequately reduce the smolt mortality associated with the descending arm of the relationship described by the ISAB flow-survival model. The technical staff's report provided a diverse set of estimates and relationships to support their positions. A well-designed, well-executed field study might shed additional light on this issue. The ISAB called for such a study in which smolt survival would be estimated under different load-following release schedules, but no formal proposal has apparently been submitted to solicit funds for such a study.

Delayed Effects Associated with Migratory Delay

There is another important dimension of the relationships between migration speed and rates of juvenile salmon survival. Extended migration travel times may cause delayed effects that could impair survival of smolts in the Columbia River estuary and after seawater entry. This hypothesis asserts that preimpoundment timing of seawater entry was synchronized with a "biological window." Extended migration travel times associated with impoundments and reduced velocities have disrupted the natural timing of ocean entry, potentially placing smolts at a disadvantage. This theoretical window has two aspects: the ecological/environmental condition of estuarine and marine waters, and the physiological condition of smolts at seawater entry.

In the late 1990s the concept of extra mortality first arose during the Plan for Analyzing and Testing Hypotheses regional modeling process. Briefly, during life cycle model analyses, total mortality exceeded that either estimated or assumed for the various individual life freshwater stages. From this modeling exercise emerged the theory that some extra or delayed effect associated with certain lifestage experiences resulted in the unexplainably low rates of salmon survival from egg through adult return. Various hypotheses, such as passage through dams and shifts in climate, were offered to explain the key driving mechanisms. Extinction risk analyses conducted in the 2000 NMFS Biological Opinion were particularly sensitive to the existence, magnitude, and persistence of this hypothetical effect.

Recent research offers additional information. Congleton et al. (2002), for example, studied changes in the condition of yearling Chinook salmon migrating from Lower Granite Dam to Bonneville Dam (1998 to 2002). In all years, body lipid and protein masses decreased significantly and with increasing travel time. The relevance of this finding is that it implies that slower migration forces juveniles to tap caloric reserves beyond normal levels. Such a tax on body reserves could thus compromise smolt performance in seawater. Although survival rates of returning adult have not yet been demonstrated to be linked to this smolt condition, these investigations suggest that salmon physiology is being compromised.

Transportation of Smolts and Delayed Effects

To expedite juvenile migration travel times through the hydropower system, smolts can be transported at dam sites that are equipped with smolt bypass/collector systems and transportation facilities. These sites include three dams on the lower Snake River (Lower Granite, Little Goose, and Lower Monumental) and McNary Dam on the Columbia River. Fish can be intercepted at these dams and transported via barge or occasionally truck to release sites downstream from Bonneville Dam. These smolts avoid in-river hazards. Even so, there is ample evidence that delayed effects on salmon attend this passage option (e.g., Giorgi et al., 2002). The magnitude and variability of these delayed effects have been identified as additional critical uncertainties within flow-survival relationships (NMFS, 2000). If the delayed effects to salmon resulting from transport around dams are not too severe, these types of transportation could be beneficial. If the effects of transport are pronounced, however, the passage strategy can put endangered stocks at risk. NOAA Fisheries is currently engaged in a multiyear research effort to help reduce mortality rates for key salmon populations in the Snake-Columbia river system associated with this type of transport.

WATER TEMPERATURE AND FLOW MANAGEMENT

Water temperature is an important factor in the life history of Pacific salmon, as it affects the rate of embryo development, juvenile growth rates, metabolic processes, and the timing of life history events such as spawning and migration (Brannon et al., 2002). In cold, high- elevation tributaries, newly emerged salmon fry must grow through the summer to obtain sufficient size to survive the lengthy downstream migration and the estuary and nearshore marine environment, then migrate to sea as yearlings. Farther downstream in the mainstem Columbia River, emergent ocean-type fry find more moderate temperatures and sufficient growth opportunities in the first spring and summer of their lives to reach sizes adequate for estuarine and marine survival during their first year or before their first year in seawater. Water temperature regimes have changed in the Columbia River (see Chapter 3), largely because of human activities. Some salmon populations have shown some ability to adapt to altered river thermal regimes. Fall Chinook salmon, for example, recently began spawning in a formerly unused site in a Snake River tributary, the Clearwater River, because water releases from Dworshak Dam[3] warmed the Clearwater River during winter, providing a suitable environment for spawning and incubation. Similarly, releases of relatively warm water from Columbia River storage reservoirs (most importantly Grand Coulee and Chief Joseph), and operation of hydro dams downstream, have increased temperature units (TU)[4] in spawning areas between the head of McNary Dam pool and Chief Joseph Dam. Adult sockeye salmon and American shad have gradually shifted the peak

---

[3] Dworshak Dam impounds the North Fork Clearwater River just upstream from Orofino, Idaho.

[4] Each 1°C for 1 day = 1 TU. Thus, for example, over 24 hours, an incubation temperature of 4 °C equals 4 TU.

of upstream migration forward about 10 days, responding to rising Columbia River water temperature (Quinn and Adams, 1996). More adult summer steelhead have tended to move later in the year, after river temperatures have peaked (Robards and Quinn, 2002). Although some adult migration and spawning times have shifted in response to lower late-spring and summer flows and warmer river temperatures, physiological responses of adult and juvenile salmon and steelhead to temperature very likely have not (Bell, 1973; Ordal and Pacha, 1963; Reiser and Bjornn, 1979a, b). High water temperatures delay the upstream migration of adult salmonids (Bjornn and Peery, 1992; Hallock et al., 1970; Major and Mighell, 1966). For example, Chinook salmon slow their movement when water temperatures approach 21° C or above (Bell, 1991; McCullough, 1999), a level already common in the Columbia River in summer (see Figure 3.8). Steelhead appear to delay migration when water temperatures exceed 21° to 22°C (Bjornn and Peery, 1992).

Clearly-defined thresholds that affect salmon behavior are difficult to identify. For example, not all Chinook salmon completely stop moving when water temperatures exceed 21°C. Fish counts at Ice Harbor Dam on the Snake River between 1962 and 1992 showed that some fish continued to move when water temperature exceeded 23.3°C (Hillman et al., 2000). Increases in summer water temperatures in the mainstem Columbia River have led to more use of cool tributary refugia (e.g., Deschutes and Wind rivers) by fall Chinook (Goniea, 2002) and steelhead (High, 2002). Higher prespawning mortality rates and depletion of energy reserves can be expected in adult fish exposed to elevated water temperature during upstream migration (McCullough, 1999; Sauter et al., 2001). There do not appear to be any analyses, however, that support precise and reliable predictions of survival changes as related to water temperature.

Within the Columbia and lower Snake rivers, summer water temperatures now reach levels that clearly impose risks to juvenile salmonids. During the summer, subyearling Chinook salmon rear and migrate downstream when river temperatures exceed 20°C (Giorgi and Schlecte, 1997). Temperature tolerance for juvenile fall Chinook has been reported to range from 5.5°C to 20°C (Groves, 1993). The young fish use more energy at high temperature, requiring either higher daily rations (that may not be available) or the consumption of stored energy.

Growth tends to decrease as water temperature approaches 19° to 20°C, which in turn can reduce the size of subyearlings at seawater entry. Disease incidence also increases with rising temperatures.

Water temperature is also an important factor affecting predation-related juvenile salmon mortality rates. For example, Vigg and Burley (1991) developed a model which suggests that a decrease in water temperature from 21.5°C to 17°C could reduce the number of prey consumed by a northern pikeminnow from seven to four per day. This suggests that water temperature regulation measures that reduced Snake River water temperatures could indirectly and locally enhance survival prospects of juvenile fall Chinook. High water temperatures during the latter part of the spring migration of smolts pose physiological threats, especially to steelhead. As previously explained, the smoltification process involves a change in physical appearance as parr become leaner and turn a silvery color. During this process, smolts become physiologically more tolerant of saltwater. Smoltification continues during the seaward migration (Beeman et al., 1995; Zaugg, 1987). Higher temperatures during downstream migration, however, can impede the smoltification process such that fish are prevented from reaching the sea.

An appropriate temperature threshold, above which smoltification is inhibited, appears to lie between 12° to 13°C (Adams et al., 1973; Zaugg et al., 1972; Zaugg and Wagner, 1973). It is not known whether actively migrating steelhead smolts that encounter temperatures greater than 14°C in the lower Columbia River, for example, would revert to parr status (for a more extensive review of temperature effects on smoltification, see *http://www.deq.state.id.us/water/suface_water/temperature/ContractorReview_EPA_DraftGuidance.pdf,* last accessed January 5, 2004). In 2001, when river flows were low and water temperatures high, survival rates of steelhead were extraordinarily low, as previously noted. And, as also noted earlier, it seems likely that the apparent "mortality" rates that year were due in part to reversion of smolts to parr status and a consequent cessation of seaward movement.

Restoration and Mitigation Measures

Flow Augmentation

In 2002, Giorgi et al. reviewed the status of flow augmentation evaluations published to date. The authors emphasized that establishing general relationships between flows and either migration speed or survival provides a rationale for entertaining flow augmentation as a strategy to improve survival. However, an evaluation of the biological benefits of providing additional water in any particular year has many facets and requires a more focused analysis. Few such detailed evaluations have been conducted. Even the 2000 NMFS Biological Opinion offered no assessment of benefits or risks associated with flow augmentation; rather, it specified volumetric (in millions of acre-feet) standards dedicated to flow augmentation and prescribed seasonal flow (in thousands of cubic feet per second, or kcfs) targets. However, no quantitative analysis describing the change in water velocity, smolt speed, or survival improvement was presented that can be attributed to the additional water provided by flow augmentation. Some studies that attempted to focus specifically on evaluating the effects of flow augmentation water delivery are discussed briefly below.

A study in the late 1990s commented on the effectiveness of flow augmentation in changing water velocity and meeting the flow targets specified in the 2000 Biological Opinion (Dreher, 1998). It was found that the volumes of water in storage reservoirs currently earmarked for flow augmentation in the Snake River (1) provide only small incremental increases in average water velocity through the hydrosystem and (2) are insufficient to meet flow targets in all years. This analysis, however, was not intended to specifically evaluate flow augmentation strategies and thus offered no insight with respect to fish responses.

The topic of summer flow augmentation has received increased attention in recent years. For example, Connor et al. (1998) conducted a study that had implications for summer flow augmentation in the Snake River. Using PIT-tagged juvenile fall Chinook that reared upstream from Lower Granite Dam, they regressed tag detection rates at the dam (survival indices) against flow and temperature separately. They found that over four years, the detection rate was positively correlated to mean summer flow and negatively correlated with maximum water temperature. They acknowledged that the predictor variables were highly correlated, limiting specific inferences regarding the effects of the individual variables. They also noted water temperatures at Lower Granite Dam dropped approximately 5° to 6°C during the period of flow augmentation from Dworshak Dam and the Hell's Canyon Complex in 1993 and 1994. They concluded that summer flow augmentation, especially cooler water released from Dworshak Reservoir, could improve survival of juvenile fall Chinook, at least to arrival at Lower Granite Dam. Connor et al. (2003) further analyzed this stock of fall Chinook salmon using PIT tag-based data for the years 1998 to 2000. Survival rates decreased as temperatures warmed and as flows decreased through the course of the summer. It was concluded that flow augmentation increased survival rates of Snake River fall Chinook salmon to the first dam they encounter.

Giorgi and Schlecte (1997) evaluated the effectiveness of flow augmentation in the Snake River for the years 1991-1995. They estimated the volume and temporal distribution of flow augmentation water delivered to the Snake River and evaluated the biological consequences to stocks listed by the Endangered Species Act. They then estimated incremental changes in water velocity and temperature that were attributable to the water delivered as flow augmentation. Using several smolt passage models, the incremental change in smolt migration speed for yearling Chinook salmon, steelhead, and fall Chinook salmon that may have resulted from flow augmentation water was estimated. It was concluded that Snake River flow augmentation increased water velocity through Lower Granite Pool an average of 3 to 13 percent during the spring. The increase was more pronounced during summers, with an increase of 5 to 38 percent change in water velocity attributable to augmentation water. Correspondingly, the change in smolt travel time predicted by the different passage models varied considerably. For example, decreases in travel time for yearling Chinook ranged from 5 to 16 percent over five years, or 0 to 5 percent depending on the passage model applied.

Temperature Manipulation

Several investigations have focused on the effectiveness of Snake River flow augmentation in reducing summer water temperature in the Lower Snake River, specifically considering the use of Dworshak Reservoir as a cold water source for decreasing water temperature in August and early September (Bennett et al., 1997; Karr et al., 1992, 1998). Karr et al. (1992) first presented results which indicated that strategic releases of outflow from Dworshak Reservoir could reduce water temperature in the Snake, at least to the vicinity of Lower Granite Dam.

Bennett et al. (1997) modeled water temperature and monitored empirical data for 1991 to 1993. They established that the Corps of Engineers model (COLTEMP) provided reliable predictions of changes in water temperature associated with flow augmentation releases upstream. The reduction in Snake River water temperature associated with cold water releases from Dworshak Reservoir was greatest at Lower Granite Dam and diminished as water moved downstream to Ice Harbor Dam. Depending on the year and base flow characteristics, the change in temperature at Lower Granite Dam typically ranged from 1° to 4°F. However, the model predicted differences as great as 6° to 8°F, which extended for a period of several weeks. Here again, prediction depended on base flows and the volume released for flow augmentation. At Ice Harbor Dam the decrease in temperature was typically small, on the order of 1 to 2°F. It was also reported that the cold water released upstream tended to sink toward the bottom of the reservoirs and mixed at the dams (Bennett et al., 1997). This suggests that deep cool water may be available as a refuge but that cooling of the entire water column cannot be achieved. Also, the extent of cooling decreases in the lower reaches of the river. Biological information has not yet been integrated with this or similar evaluations.

Benefits and Risks to Other Species

Water releases from storage reservoirs to increase mainstem flows or to reduce water temperatures alter conditions both in the storage reservoirs and in tributaries connecting with the Columbia and Snake rivers. These processes in turn have effects on resident and anadromous fish inhabiting those waters, which introduces an additional, complex facet of flow augmentation. Risks associated with flow augmentation were addressed by the Independent Science Advisory Board's publication *Return to the River*, which expressed concerns regarding risks associated with summer flow augmentation, in particular (ISAB, 1996):

> Underscoring these substantial uncertainties in flow augmentation rationale is the fact that summer drawdowns in upstream storage reservoirs, for example Hungry Horse Reservoir in Montana, to accomplish summer smolt flushing in the lower Columbia River has direct and potentially negative implications for nutrient mass balance and food web productivity in Flathead Lake, located downstream from Hungry Horse.

The issue involves balancing expected benefits to anadromous fish with ecosystem functions and potential risks to other species. There is clearly a complex array of water management activities in the Columbia River basin today, and arriving at an appropriate balance among competing and complementary strategies is a venture that contains many considerations and uncertainties.

Flow Management and the Estuary

The ISAB (1996) stressed the importance of the estuary as a key regulator of overall survival and annual variation in abundance of salmon. The estuary (and nearshore Columbia plume and its interface with seawater) provides a physiological transition zone, potential refuge from predators, and forage (Simenstad et al., 1982). Rapid growth of juvenile salmon in this transition zone is important, as increased size lessens vulnerability to predation in this environment. For example, in the lower Sacramento River, the primary floodplain area provides better rearing and migration habitat for juvenile Chinook salmon than provided by adjacent river channels (Sommer et al., 2001). Anthropogenic effects on estuarine and plume dynamics derive from estuarine alterations such as diking and filling, and from flow and water quality alterations upstream (e.g., reductions in turbidity;

Junge and Oakley, 1966).

The Columbia River estuary has changed greatly since the early 1800s. Total volume of the estuary has declined by about 12 percent since 1868, and diking and filling have converted 40 percent of the original floodplain to various human uses (Sherwood et al., 1990). The annual spring freshet has been greatly diminished, thereby reducing organic and sediment inputs. The standing crop of organisms that feed on macrodetritus is only about one-twelfth as great as it once was (ibid.). The Northwest Power Planning Council's ISAB (1996) assumed that a reduction in the food web supported by phytoplankton macrodetritus has negatively affected salmon. Changes in food web production have resulted in a more favorable environment for herring, smelt, and shad. Estuarine degradation and potential mitigation are further discussed in Bottom et al. (2002), Jay and Naik (2000), and Kukulka and Jay (2003). Hatchery-produced salmon and steelhead now pass through the estuary in large quantities, in temporal patterns dissimilar to historical patterns of the passage of wild fish. Effects of these large releases on estuarine ecology are not fully understood and quantified. Nonetheless, they are likely to negatively affect wild anadromous fish because of the diminished ecological opportunities offered by a smaller estuary that has experienced pronounced hydrologic and related changes.

Tributary and Riparian Issues

Potential exists to increase salmon stocks in the Columbia River system by restoring or rehabilitating riparian vegetation that has been altered by overgrazing, timbering, mining, and clearing for agriculture (Maloney et al., 1999; Meehan, 1991). For example, approximately 88 percent of the original presettlement forests occupying the floodplain of the Willamette River (a major tributary of the Columbia) have been removed (NRC, 2002a). A pristine riparian zone, unaltered by human activities, enhanced salmon spawning and rearing by)1) shading the stream and maintaining low water temperatures, (2) contributing coarse woody debris to provide cover and in-stream habitat heterogeneity, (3) filtering sediment and pollutants from runoff waters, and (4) producing many forms of organic matter to support stream productivity (Clinton et al., 2002; McIntosh et al.,

1994; Naiman et al., 1992). Returning adult salmon themselves contribute to riparian zone and stream productivity by transporting marine-derived nutrients to their spawning grounds (Schindler et al., 2003).

SUMMARY

Columbia River salmon are anadromous and are affected by environmental conditions and variability not only within the Columbia River basin but also by conditions in the northern Pacific Ocean. Columbia River basin salmon have been in a general state of decline for decades, with these declines being driven by a variety of environmental changes. There have been departures from this long-term trend, the most recent being an increase in the returns of (mainly hatchery-reared) Chinook salmon in 2002 and 2003. This increase has generally been attributed to favorable ocean conditions. Although a positive development, these increased numbers still fall well short of what was once the world's premier salmon fishery. Despite some recent increases in returns, there is little disagreement on general long-term declining trends, which have resulted in many wild salmon species being listed as threatened or endangered under the Endangered Species Act.

This report reviews the implications for salmon survival of a specific and relatively (compared to the magnitude of the Columbia River) small range of proposed water withdrawals that would further reduce river flows. Precise and credible forecasts of specific biological or ecological outcomes of these withdrawals (or almost any given range of specific proposed diversions) are beyond current scientific capabilities and knowledge. But as pointed out in Chapter 3, impacts of water withdrawals from the Columbia River on salmon survival rates vary according to seasonality of withdrawals. During periods of high base flows, and assuming that future seasonality of water withdrawals does not change, the upper end of the magnitude of water permit applications being considered in this report (1.3 million acre-feet) will have only minimal effects during periods of low water demand and low withdrawal rates. However, during the summer months of high water demand, the upper range of the prospective withdrawals considered in this report would decrease flows in the

Columbia River considerably, especially if these additional withdrawals were diverted during lower-than-average flows during July and August. Moreover, cumulative effects of individual withdrawals eventually result in important thresholds being crossed and with resulting deleterious effects on salmon. Trends such as likely future climate warming across the Columbia River basin; potential additional withdrawals from the Columbia Basin Project, upper basin states, provinces, and tribal reservations; degraded water quality, and periodic poor ocean conditions for salmon all point to additional risks in maintaining viable Columbia River salmon populations. The coincidence of more than one or all these unfavorable trends could have serious negative consequences for Columbia River salmonids. Given the current setting and likely future trends, additional withdrawals from the Columbia River during the summer months of high water demand and during low-flow years will pose substantial additional risks to salmon survival. These risks vary across salmon stocks, with stocks that inhabit the Columbia mainstem during low-flow periods exposed to greater risks. These greater risks to salmon survival should be carefully considered in decisions regarding potential future Columbia River withdrawals during low flows.

Selecting the "best" model of salmon-environmental relationships was neither part of this study nor critical to its completion. Analyses and models presented by several expert scientists during open public meetings in the course of this study were used as background information for considering the degree to which additional water diversions, as well as changes to the river's thermal regime, may pose increased risks to the survival of endangered fish species. This information, along with the large body of scientific evaluations of Columbia River salmon and their habitat, portrays a complex and only partially understood picture of the relative influences of many different environmental variables on salmon survival rates. Efforts to identify whether water velocity, temperature, or some other variable(s) are among the more important factors affecting juvenile salmon survival rates, or identifying critical thresholds associated with these variables, are therefore problematic. **Within the body of scientific literature reviewed as part of this study, the relative importance of various environmental variables on smolt survival is not clearly established. When river flows become critically low or water temperatures excessively high, how-**

ever, pronounced changes in salmon migratory behavior and lower survival rates are expected.

The issue of water use permitting decisions is controversial, as these decisions have important environmental, economic, and social implications. Instituting water use permit and extraction policies that vary according to season and river flows will require greater flexibility in these institutions than currently exists. This greater flexibility will be necessary, however, if risks to salmon survival are to be better managed and if salmon management is to move toward more adaptive regimes than used in the past. In addition to greater institutional flexibility, additional cooperation across the entire Columbia River basin appears necessary to better manage risks to salmon. For example, if the State of Washington and its water users exercise caution and restraint in considering the issue of additional water withdrawal permits for low-flow periods, the benefits of any measures will be decreased or negated if other entities in the basin do not adhere to similar practices. The following chapter reviews efforts at cooperation across the Columbia River basin and identifies some of the limits of and lessons from these efforts and what they bode for future cooperative regimes across the basin.

5

Water Laws and Institutions

INTRODUCTION

In addition to impoundments, dams, diversion structures, and numerous environmental factors, the migratory and life cycle patterns of Columbia River salmon are affected by a sophisticated legal, institutional, and decision-making framework. This framework reflects the jurisdictional complexity of the Columbia River basin and a patchwork of treaties, legislative enactments, executive directives, and court rulings. The Columbia River is one of North America's most jurisdictionally complex rivers. The river's basin extends into two countries, seven states, and hundreds of other governmental subdivisions. The basin is home to 13 Indian tribes, and eight federal agencies have water-related resource responsibilities in the basin (Blumm and Swift, 1997). Finally, salmon that are reared and that spawn in the basin spend a substantial portion of their lives traversing the international waters of the Pacific Ocean.

This chapter discusses some of the laws and institutions that govern water resources management decisions in the Columbia River basin. It is not meant to comprehensively review and interpret all laws and policies that guide river management but rather is designed to illustrate the many sources of risk that affect decisions in permitting additional water uses in the stretch of the Columbia River that flows within the State of Washington downstream from the Canada-U.S. border. This is consistent with this report's emphasis on the implications of water withdrawals from the mainstem Columbia River in the State of Washington (the "middle reach" of the Columbia). The key themes of this chapter are the prospects of additional diversions upstream of the Columbia middle reach in Washington and the challenges that additional withdrawals will pose for the existing legal and institutional framework in the state and across the river basin.

INTERNATIONAL OBLIGATIONS

Pacific Salmon Treaty

The Pacific Salmon Treaty (16 U.S.C. §§ 3631-3644, March 15, 1985) was concluded in 1984 and ratified by Canada and the United States in 1985. The treaty grants each country four commissioners. The U.S. delegation is composed of one commissioner from Alaska, one commissioner representing the states of Oregon and Washington, one commissioner representing the 24 tribes, and one nonvoting federal commissioner. Representatives from these governments also serve on several subsidiary panels. The treaty's goal is "coordinated management of Pacific salmon throughout their range to ensure sustainable fisheries and maximize long-term benefits to the parties" (Waldeck and Buck, 1999). Under the 1999 agreement, the parties agree to an "abundance-based," or supply-side, approach to management and harvest. The 1999 agreement emphasizes the importance of habitat in achieving treaty goals. The parties pledge "[t]o use their best efforts, consistent with applicable law, to: (a) protect and restore habitat so as to promote safe passage of adult and juvenile salmon and achieve high levels of natural production, (b) maintain and, as needed, improve safe passage of salmon to and from their natal streams, and (c) maintain adequate water quality and quantity."[1]

Significance for the Columbia River middle reach: The Pacific Salmon Treaty, with its focus on salmon harvest limits, does not impose any direct regulation on water management in the river's middle reach. However, through its ratification of the treaty, the U.S. federal government defines a foreign policy objective of sustaining the salmon fishery and protecting and improving salmon habitat in and passage through inland waters. Increased consumptive diversions in the Columbia River's middle reach, with possible habitat modifications, might produce results contrary to these foreign policy goals.

---

[1] Att. E, Habitat and Restoration, Annex 4 to Treaty Between the Government of Canada and the Government of the United States of America Concerning Pacific Salmon (*http://www.psc.org/treaty*).

Columbia River Treaty

The Columbia River Treaty[2] was signed in 1961 by representatives of Canada and the United States and was ratified by the two governments by 1964. The treaty provided for the construction of four upper Columbia River basin storage dams: Duncan (1967), Keenlyside (1968), and Mica (1973), all in Canada, and Libby in Montana (1973). These dams provided flood control and increased hydroelectric power generating potential in both countries. The 60-year treaty coordinates binational flood control and electrical energy production in the Columbia River basin. Pursuant to the treaty, Canada stores 15,500,000 acre-feet of water in upstream storage reservoirs. In return, Canada received one-half of the additional power generated at downstream U.S. power plants by this 15,500,000 acre-feet of water. As each Canadian dam was completed, hydropower benefits that were generated downstream (and owned by the province of British Columbia) were sold to a group of U.S. utilities for a 30-year period. The first 30-year contracts began to expire in 1998. British Columbia is now receiving the sales revenues of those downstream benefits for the remaining 30 years of the treaty. For 2000 to 2001, British Columbia received $632 million as its share of hydropower revenues. Some of this money is assigned to a Canadian Columbia Basin Trust.

The treaty provides for an "entity" from each country. The U.S. entity refers to the administrator of the Bonneville Power Administration and the division engineer of the Corps of Engineers North Pacific Division, who together implement the operating arrangements necessary to enforce the Columbia River Treaty. For Canada, under a separate British Columbia-Canada agreement, British Columbia Hydro is designated as the Canadian entity responsible for executing Canadian obligations under the treaty, including construction of the three Canadian dams.

The treaty has several important water rights features. Canada has certain rights to divert water from the Kootenay River into the headwaters of the Columbia. Between the 20th and 60th years of the treaty, this may be as much as 1,500,000 acre-feet

---

[2] Treaty with Canada Relating to Cooperative Development of the Water Resources of the Columbia River Basin, 15 U.S.T. & O.I.A., vol. 2, T.A.A.S. No. 5638. See also Johnson, The Canada-United States Controversy Over the Columbia River, 41 Wash. L. Rev. 676 (1966).

per year. For 40 years after the treaty expires, Canada can continue to divert unspecified amounts of water from the Kootenay River into the Columbia, so long as Kootenay River flows at the border are 2,500 cubic feet per second or the natural flow. The treaty is not a general apportionment of Columbia River waters. Canada pledges not to divert water in a way that alters the flow of water crossing the boundary, but an exception is made for consumptive uses. This restraint is designed to prevent trans-basin diversions, such as into the Fraser River (Canada's controversial proposed project that led to the 1961 treaty).

Significance for the Columbia River middle reach: So long as the level of hydropower production under the treaty is maintained, there should be no significant changes to water availability in the Columbia River's middle reach. Without U.S. consent, Canadian transfers out of the river's basin are prohibited. The water transfer between the Kootenay River (which ultimately flows into the Columbia) into the headwaters of the Columbia generally has limited hydrological implications for the Columbia River middle reach. The treaty is not an apportionment of the river between the two countries, however, and other international law principles, such as the Boundary Waters Treaty, must be considered.

Boundary Waters Treaty

The principal water management and sharing mechanism between Canada and the United States is the Boundary Waters Treaty.[3] Ratified in 1909, it created the bilateral International Joint Commission (IJC) to address disputes. Several provisions of the treaty address the apportionment of boundary waters between Canada and the U.S. For example, under Article I, each country is entitled to "exclusive jurisdiction and control over the waters" on its side of the border. Several other provisions dampen this exclusive jurisdiction rule. Under Article II, a party injured by an upstream diversion in the other country has the same legal rights as a resident of the upstream nation. Under Article VIII, each nation has "equal and similar rights in the use" of

---

[3] Treaty Relating to Boundary Waters and Boundary Questions, 36 Stat. 3488 (1909).

boundary waters. These somewhat contradictory provisions may result in adoption of an equitable apportionment or an equal division of boundary waters (Tarlock, 2000).

There is a possibility that additional Columbia River water could be developed by Canada, and it is unclear what the legal implications would be for water uses in the river's middle reach. In the case of increased Canadian diversions, a downstream water user in the State of Washington would have the same rights to contest the diversion as a Canadian resident; but application of the equitable apportionment principle usually means (at least in U.S. jurisprudence) that actual water uses within a state must not exceed that state's equitable share of the interjurisdictional water source. As a practical matter, injury to Columbia River middle reach users as the result of increased Canadian diversions would be processed through time-consuming IJC procedures. The U.S. State Department controls how such cases are presented.

Significance for the Columbia River middle reach: Current population growth rates in British Columbia suggest that increased Columbia diversions are likely, which will reduce downstream flows. Additionally, Canada likely has an unquantified but, for purposes of prior appropriation in Washington, a senior claim based on its equitable interest in the river. Canadian development will thus result in incrementally less water in the river. Additional U.S. water diversions in the river's middle reach may remain subject to additional Canadian development, the latter of which would be entitled to priority. This does not consider any water-related claims of indigenous people north of the 49th parallel.

INTERSTATE COMPACTS

Attempted Columbia River Basin Compact

From 1950 to 1968, the states of Montana, Idaho, Oregon, and Washington attempted the negotiation of a Columbia River Basin Compact (Nevada, Utah, and Wyoming were minor participants). Although much of the discussion concerned upper- and lower-basin allocations of water, the debate really focused on the rivalry between public and private hydropower generation. The movement for public power in the Northwest had resulted in a proposed Columbia Valley Authority for the region in the late 1940s, but private power interests held a political advantage during the Eisenhower administration. Upper-basin states such as Montana, with a history of private power development, supported a compact as a means of promoting private power interests. Although a compact was signed by the compact commissioners and approved by Congress, it ultimately failed when the Oregon and Washington legislatures failed to ratify the document. The central compact feature was a trade of upper-basin storage for hydropower. The upper-basin states would have allowed the construction of larger reservoirs in exchange for a share of future hydropower production and a guarantee that much of their future consumptive water needs would prevail over lower-basin instream uses.

Columbia River Compact

Although the quest to establish a basinwide water quantification compact was unsuccessful, a compact was reached concerning commercial and recreational fisheries. The Columbia River Compact provides authority to adopt seasons and rules for Columbia River commercial fisheries. Compact administration is by the Oregon and Washington agency directors, or their delegates, acting on behalf of the Oregon Fish and Wildlife Commission and the Washington Fish and Wildlife Commission. The basic text of the compact is as follows:

> All laws and regulations now existing, or which may be necessary for regulating, protecting or preserving fish in the waters of the Columbia River, over which the States of Oregon and Washington have concurrent jurisdiction, or any other waters within either of said states, which would affect the concurrent jurisdiction, shall be made, changed, altered and amended in whole or in part, only with the mutual consent and approbation of both states. (Oregon Rev. Stat. § 507.010).

When addressing commercial seasons for salmon, steelhead, and sturgeon, the compact considers the effect of the commercial fishery on escapement, treaty rights, and sport fisheries, as well as the impact on species listed under the Endangered Species Act. Although the compact does not provide authority to adopt sport fishing seasons or rules, it does address the allocation of limited resources among users.

Significance for the Columbia River middle reach: The compact is designed to regulate commercial fishing, but the language concerning laws or regulations "necessary for regulating, protecting or preserving fish" has the judicially untested potential of requiring greater collaboration between Washington and Oregon on anadromous fish issues. Water rights permitting decisions, unless they require a new statute or rule, do not appear to be affected by this treaty.

Northwest Power Act and the Northwest Power and Conservation Council

Throughout the twentieth century, growth and demand for electric power, irrigated farmland, and flood control in the Pacific Northwest were met by increasingly large water storage structures. Until the 1970s, power and other services provided by the system were generally viewed as beneficial and essential to the region's growth. By then, however, the benefits of the system were increasingly challenged, as environmental, economic, and social costs of construction were raising questions and doubt. In 1980, Congress passed the Pacific Northwest Electric Power Planning and Conservation Act, which authorized the states of Idaho, Montana, Oregon, and Washington to create the Northwest Power Planning Council. Renamed the Northwest Power and Conservation Council (NPCC) in 2003, the council consists of eight board members, two appointed by the governor of each state. The act established two objectives for the council: (1) to forecast power demands in the region, and (2) to plan for mitigation associated with the FCRPS. The act also directed the council to pay particular attention to information provided by Native American tribes. The council is responsible for mitigating the impacts of hydroelectric power dams and their operations on all fish and wildlife in the Columbia River basin, including endangered species, through a program of enhancement and protection. The council is intended to be a broker among many contending interests including agencies, tribes, electric utilities, and environmental and business interests. The fish and wildlife program of the council directs the expenditure of hundreds of millions of dollars per year of federal Bonneville Power Administration revenues intended to mitigate damages to fisheries.

Among the key features of today's NPCC is its authority to guide the actions of federal agencies. The Bonneville Power Administration, for example, is required to ensure that its actions are consistent with NPCC plans and initiatives, and other federal agencies are required to consider the council's programs at each stage of the decision-making process. Flows of information for decision making within the council are complex, as they include large numbers of committees and advisory bodies. The council seeks input from research projects, agency initiatives, and networking workshops. Information is provided from a variety of stakeholder and community sources through public hearings, outreach, and public advisory groups.

In 2000 the Northwest Power Planning Council established a geographically based plan for implementation. The program is to be implemented through subbasin plans developed locally in more than 50 tributary subbasins of the Columbia River and amended into the council's program. The efficacy of this grassroots implementation strategy remains to be seen. The complex organizational arrangements engaging large numbers of professional and public advisors serve to spread the risks of failure over large numbers of participants as well as co-opt potential critics. In some sense, issues are "domesticated rather than addressed, and hard problems are removed from the day to day decision space" (Rayner et al., 2000). Although problems may not be fully resolved, such strategies allow for additional time and resources in which to search for alternative solutions and in which public tastes and values may undergo changes.

INTERSTATE APPORTIONMENT

Three traditional methods have been used to resolve interstate water disputes. One approach for addressing regional intergovernmental water disputes is the interstate compact. Compacts are specifically authorized by the U.S. Constitution and were first used for resolving boundary conflicts. Compacts require congressional authorization, either before or after the agreement is reached; and, once a compact has been approved by Congress, it has the statute of federal law under what is known as the Law of the Union doctrine. The first water quantification compact in the United States, allocating water between the upper and lower basins of the Colorado River, was negotiated and ratified in the 1920s. Since congressional approval of this initial compact, over 20 other water compacts have been negotiated throughout the United States. Since the 1980s, several states and tribes have negotiated congressionally approved compacts or other agreements determining tribal reserved water rights.

A second method for addressing intergovernmental natural resource disputes is federal legislation. In interstate conflicts over water, this method, known as a congressional apportionment, has rarely been used: once to allocate water among Colorado River basin states and, implicitly, in water quality disputes in the Great Lakes. Although federal legislation could provide a comprehensive water allocation agreement for the Columbia River basin, members of Congress are rarely able to reach agreement among themselves about divisive regional issues. Many of them also believe these disputes are better left to local resolution.

A third traditional approach to addressing interstate water disputes involves litigation. For water-related disputes among states, the Constitution provides that the U.S. Supreme Court has original and exclusive jurisdiction to hear these cases. If a dispute involves the interpretation or enforcement of an existing interstate compact, the Supreme Court usually will look to that document for the principles necessary to resolve the matter. If no compact exists or an existing compact does not address the dispute, the Supreme Court may apply a set of federal common law rules to reach an equitable apportionment of the water resources of the water body. Because these original jurisdiction cases require a factual record, they are usually referred first to a court-appointed special master who holds hearings and submits a proposed resolution of the case to the Supreme Court for its review.

The utility of using these three traditional methods to resolve complex water quality disputes or regional endangered species problems generally has not been tested. One exception is the Delaware River Basin Compact, approved in 1961 by Delaware,

New Jersey, New York, Pennsylvania, and the United States. This state-federal compact is governed by a commission of the governors from the four states and a federal representative appointed by the president. The compact's most distinctive feature is its requirement that the commission is charged to develop and implement a comprehensive basin plan. The compact also gives the commission licensing authority by providing that "no project having a substantial effect on the water resources of the basin shall hereafter be undertaken unless it shall have been first submitted to and approved by the Commission." The commission must approve the proposed project if it "would not substantially impair or conflict with the comprehensive plan." The Delaware River Basin Compact is similar to the 1980 Pacific Northwest Electric Power Planning and Conservation Act in that it also created a four-state commission, which also addressed multiple resources and required the development of a regional energy plan (which is presumptively binding on federal agencies). Another federal-state arrangement for coordinating multiple jurisdictions in a U.S. interstate river basin is in the Susquehanna River basin (*http://www.srbc.net/*, accessed February 17, 2004).

More recently, governments sharing regional water bodies have used less formal, and more flexible arrangements to address interjurisdictional water issues. These include the Enlibra conflict resolution principles endorsed by the Western Governors Association, statements of guiding management principles such as the Great Lakes Charter, multifaceted state-federal agreements (e.g., California's CalFed Bay-Delta program), and drought or water banks such as those used in Idaho and the lower Colorado River. All of these arrangements may be useful in increasing the flexibility of traditional water management regimes (e.g., the doctrine of prior appropriation) across the Columbia River basin and may be helpful in addressing existing and emerging water allocation issues.

NATIVE AMERICAN WATER AND FISHERIES RIGHTS

Legal Basis

Indian claims to water and fish are usually based on the federal organic document that established a reservation of land for the tribe: a treaty, statute, or presidential executive order. These documents sometimes make explicit statements concerning these resources. They might indicate, for instance, that the tribe has reserved to itself an existing fishery right. These documents are often silent about tribal resources, but the courts have read an "implied" reservation into these agreements or documents, recognizing that neither the tribe nor Congress would have intended a reservation of land without water. Finally, tribes may assert aboriginal rights independent of any document. These claims are based on extended exclusive occupancy of land before forceful removal (Cohen and Strickland, 1982). The Pacific Northwest has produced many judicial opinions that have been hallmarks in the development of Indian law as it pertains to resource management. These cases often involved (and still do) the intersection of fisheries and water resource issues. The foundational legal case in this realm is *United States v. Winans* (198 U.S. 371, 1905), as it serves as the common spring for the law of Indian fisheries and the reserved water rights doctrine (Box 5-1 lists the Columbia River basin tribes).

United States v. Winans (1905)

This U.S. Supreme Court decision announced reserved right principles (that would be further developed in the *Winters* case in 1908) that held that the tribes' rights of taking fish at all usual and accustomed places in common with the citizens of the territory of Washington, and the right of erecting temporary buildings for curing them, were reserved to the Yakama Nation in the treaty of 1859. The court ruled that this was not a grant of right but a reservation of rights already possessed and not granted away. The rights so reserved imposed a servitude on the entire land relinquished to the United States under the treaty and which, as was intended to be, was continuing against the United States and its grantees as well as against the state and its grantees.

United States v. Winters (1908)

In the 1908 case of *United States v. Winters* [207 U.S. 564

BOX 5-1
Columbia River Basin Tribes and Reservations

Burns Paiute Tribe (Oregon)—3,000 members; 770 acres of trust land acquired in 1935 to reestablish reservation; 11,000 acres of allotment land owned by tribal members.

Coeur d'Alene Tribe (Idaho)—1,700 members; 345,000-acre reservation; rights based on treaties as early as 1873.

Confederated Salish and Kootenai Tribes of the Flathead Reservation (Montana)–6,900 members; 1,300,000 acre reservation; assert rights based on 1855 Treaty of Hellgate.

Confederated Tribes of the Colville Reservation (Washington)—8,400 enrolled members; 1,400,000 acre reservation; rights based on 1872 Executive Order and other agreements with the U.S. government (1892, 1905).

Confederated Tribes of the Umatilla Indian Reservation (Oregon)—2,174 enrolled members; 180,441-acre reservation; rights based on 1855 treaty.

Confederated Tribes of the Warm Springs Indian Reservation (Oregon)—3,916 enrolled members; 650,000-acre reservation; rights based on 1855 treaty and federal court cases.

Kalispel Tribe of Indians (Washington)—280 enrolled members; 4,600-acre reservation; rights based on 1914 executive order.

Kootenai Tribe (Idaho)—67 members as of 1974; tribal members accepted 12.5 acres but do not consider it to be a final settlement.

Nez Perce Tribe (Idaho)—3,200 members; 770,453-acre reservation; rights based on treaties of 1855 and 1863 and federal court decisions.

Shoshone-Bannock Tribes of the Fort Hall Reservation (Idaho)—4,291 members; 544,000-acre reservation; rights based on 1867 executive order.

Shoshone-Paiute Tribes of the Duck Valley Reservation (Nevada)—1,818 members; 289,820-acre reservation; rights based on 1863 treaty, 1877 executive order, and other statutory additions to reservation.

Spokane Tribe of Indians (Washington)—100,000 acres held in trust; 57,370 additional acres held as allotments, deeded fee land, other government lands; rights based on 1880 executive order.

Yakama Nation (Washington)—9,092 members; 1,390,000 acre reservation; rights based on 1855 treaty.

(1908)], which arose on the Milk River in the State of Montana, the Supreme Court recognized that the reservation system had been established in an effort to transform tribes into agrarian societies. The court ruled that Congress reserved, by implication, sufficient water to serve the needs of the reservation with a priority extending back to the date the reservation was established. In some cases these *federally reserved* water rights are claimed as aboriginal, based on historic use, with a priority date of "time immemorial." Since Indian fishing and water rights claims are senior to most non-Indian uses, there has been a slow but continuing effort to quantify these treaty, or reserved, water rights. Quantification can be by litigation, compacts or settlements, or congressional legislation. All tribes with trust status reservations within the Columbia River basin and its tributaries potentially have treaty-based or reserved water rights claims. Quantification of nonfishing claims has been based on the practicably irrigable acreage standard (see *Arizona v. California,* 1963). The Arizona Supreme Court, however, recently utilized a "permanent homeland standard" in the Gila River adjudication that may stimulate further discussion of appropriate quantification methods.

Indian Fisheries Cases in Washington

Under the *Winans* case, tribes may reserve by treaty the right to hunt or fish off-reservation. This legal principle is at the heart of lengthy litigation in state and federal courts in Washington State.

Puyallup Cases

The chronicle of litigation begins in 1968 with *Puyallup Tribe v. Department of Game* [*Puyallup I,* (391 U.S. 392 (1968)], decided by the U.S. Supreme Court. The tribe had entered into a Stevens treaty[4] in 1854 that stated: "The right of taking fish, at all usual and accustomed grounds and stations, is further secured to said Indians, in common with all citizens of the

---

[4] The reservations and off-reservation rights of Columbia River basin tribes are established by a series of "Stevens Treaties" named after a former governor of Washington who negotiated with the tribes.

Territory."⁵ Washington State attempted to prohibit tribal members, when fishing off-reservation, from using nets. The Supreme Court upheld the state's qualified authority to regulate the tribe's fishing right: "But the manner of fishing, the size of the take, the restriction of commercial fishing, and the like may be regulated by the State in the interests of conservation, provided the regulation meets appropriate standards and does not discriminate against the Indians" (391 U.S. at 398). In a later case, "appropriate standards" were defined to mean a reasonable and necessary conservation measure, the applicability of which to Indians "is necessary in the interests of conservation" [*Antoine v. Washington*, 420 U.S. 194, 207, (1975)].

Soon thereafter, the State of Washington allowed tribal members to use nets for salmon but not for steelhead. The tribe argued that this restriction resulted in assigning the entire run to non-Indian sports fishermen. When this restriction was reviewed by the U.S. Supreme Court in *Puyallup II* [*Department of Game v. Puyallup Tribe*, 414 U.S. 44 (1973), *Puyallup II*], the justices indicated that regulation discriminated against the tribe and violated *Puyallup I*. The court suggested that some accommodation between Indian and non-Indian uses had to be found; but, if necessary, a nondiscriminatory fishing ban to save steelhead could be applied to Indians. In a third round of litigation, the state allowed the Indians to net steelhead but limited their share to 45 percent of the natural run. Contemporaneously, many tribal "usual and accustomed" fishing locations were determined to be within reservation boundaries, although still on non-Indian land. The tribe challenged this state limitation as well, particularly as applied to on-reservation locations. In *Puyallup Tribe, Inc. v. Department of Game* (433 U.S. 165, 1977, *Puyallup III*), the Supreme Court upheld the state regulation and allowed it to be applied to on-reservation fishing so as to prevent the tribe from taking an unlimited amount of fish to the detriment of non-Indian fishermen.

Boldt Litigation

While the *Puyallup* litigation was pending, the United States

---

⁵ Treaty with the Nisqually and Other Indians, art. III, 10 Stat. 1132, 1133 (1854).

filed suit in federal district court in 1970 on behalf of seven Washington-based tribes that asserted fishing rights based on the same Stephens treaty language. On February 12, 1974, Judge George Boldt ruled in *United States v. Washington* [384 F. Supp. 312 (W.D. Wash. 1974), aff'd, 520 F.2d 676 (9th Cir. 1975), cert. denied, 423 U.S. 1086 (1976)] that the tribes had a right to fish at their accustomed places and to secure roughly half of the annual catch. More specifically, the district court held that the Indians were entitled to a 45 to 50 percent share of the harvestable fish that would at some point pass through recognized tribal fishing grounds in a defined area of Washington, to be calculated on a river-by-river, run-by-run basis, subject to certain adjustments. With slight modification, the U.S. Court of Appeals for the Ninth Circuit affirmed, and the U.S. Supreme Court declined review. In a later decision, Judge Boldt declined to extend federal recognition or enforce treaty rights for certain landless tribes (the Duwamish, Samish, Snoqualmie, and Steilacoom). Although the district court ordered the state fisheries department to adopt regulations protecting tribal fishing rights, a state court action resulted in the Washington Supreme Court holding that state agencies could not comply with the federal court injunction. The state court ruled that the treaty conferred on the Indians no greater right than that enjoyed by non-Indians. To rule otherwise, in the court's view, would violate the Equal Protection Clause (*Puget Sound Gillnetters Ass'n v. Moos*, 565 P.2d 1151, Wash. 1977).

These various federal and state decisions were eventually all reviewed by the U.S. Supreme Court in 1979. In rejecting the ruling of the state supreme court, the court vindicated the federal district court's approach. In *Washington v. Washington State Commercial Passenger Fishing Vessel Ass'n* [443 U.S. 658 (1979)], the Supreme Court held that the treaties do not guarantee merely access to the fishing sites and an equal opportunity for Indians and non-Indians to fish, but rather secure to the Indian tribes a right to harvest a share of each run of anadromous fish that passes through tribal fishing areas. Among the more specific holdings:

• A 50 percent share of the harvestable run may be established as the ceiling for the Indian fishery. This share may be reduced when fish are not needed, for instance, if a tribe's popu-

lation has declined.
- The state has the authority to set the harvestable run for each stream in a manner that protects the sustainability of each run.
- All fish caught by treaty Indians count against the tribal share, whether caught on- or off-reservation.
- All fish caught by non-Indians count against their share, whether or not caught in state waters.
- Indians are entitled to the exclusive use of all fishing sites within reservation boundaries (Canby, 1981).

Significance for the Columbia River middle reach: The *Boldt* litigation, culminating in the 1979 U.S. Supreme Court decision, recognizes state authority to determine harvestable catch for both Indian and non-Indian fishermen. That authority, however, is tempered by the obligation to manage the resource in a manner that safeguards the sustainability of the resource. In practice, management of the fishery has become more of the collective responsibility of federal, state, and tribal fish managers. Still, the state must be cautious in its water permitting function not to affect the salmon and steelhead resource in such a way that no harvestable catch is available for treaty Indians or to take actions that are detrimental to the sustainability of existing runs.

Water Rights of Indian Reservations

As previously discussed, many of the Stephens treaties reserved tribal rights to fish on the reservations and at off reservation "usual and accustomed" sites in their treaties—the provision interpreted in the *Puyallup* and *Boldt* litigation. The total land represented by these reservations exceeds 7,000,000 acres (or roughly 11,000 square miles, about the size of the combined area of Massachusetts and Connecticut). Water rights for some of this tribal land have been adjudicated or settled. Other land may not have been reserved for agricultural purposes or may be of poor quality. If, however, irrigated agriculture was "feasible" on 25 percent of this land, and 4 acre-feet of irrigation water per acre per year were required, 7,000,000 acre-feet of water could be diverted from the Columbia River system for farming ("feasibility" of irrigation is a technical and economic concept used in defining "practical irrigable acreage"; its calculation depends on site-specific conditions and studies and, depending on assumptions, can vary widely). The following discussion broadly examines some of the larger reservations to gauge how their claims and uses might affect water availability in the Columbia River middle reach.

Yakama Nation/Yakima Indian Reservation (Washington)

The Yakima River flows from the northwest and empties into the Columbia River at Richland, Washington. Water rights established on the Yakima River affect water availability downstream on the Columbia River mainstem. The Yakima River has been the subject of the ongoing Yakima River adjudication, originally filed by the State of Washington in 1977. The water rights of the Yakama Nation have been asserted in the adjudication, and several important decisions have been reached. In November 1990 the Yakima County Superior Court granted a partial summary judgment establishing the quantity and priority of treaty-reserved water rights for irrigation within the Yakama Reservation and for fishing purposes both within the reservation and off-reservation in the Yakama Nation's "usual and accustomed" fishing area. The court determined that federal legislation passed in 1914 and subsequent federal legislative, executive, and judicial actions had reduced the amount of water claimed by the tribe under its treaty. Consequently, the court ruled that the tribal fishing right was limited to the instream flow necessary to maintain fish life in the river. The case was appealed to the Washington Supreme Court, which in April 1993 affirmed the lower court decision.

Significance for the Columbia River middle reach: The Yakima River adjudication is nearing conclusion. In addition to the quantification of the Yakama Nation's water rights, major non-Indian irrigation claims have been resolved through litigation and settlement. As a result of this adjudication, the state has acquired reasonably current and accurate information regarding the use of water on this Columbia River tributary stream. The tribal fishing right, although less than claimed, remains a "time immemorial" instream flow right that must be protected by the state in future permitting decisions.

Confederated Tribes of the Colville Reservation (Washington)

Twelve bands or tribes of indigenous people were located on land within the territory of Washington pursuant to a presidential executive order issued in April 1872. On July 2 of the same year a second presidential executive order moved the reservation and the residents to its present location on the west side of the Columbia River. Although this location originally totaled almost 3,000,000 acres, subsequent enactments reduced the acreage to the present size of 1,400,000 acres. Tribal members, however, retain hunting and fishing rights on the ceded northern half of the original reservation (*Antoine v. Washington*, 420 U.S. 194, 1975). Grand Coulee Dam and the lower part of Roosevelt Reservoir are located within the external boundaries of the reservation; the upper lake is within the ceded areas. Tribal membership is approximately 8,700, about half of whom live on or adjacent to the reservation.

The reservation was the location of the *Colville Confederated Tribes v. Walton* (1981) decision, which recognized the ability of non-Indian assignees of Indian allotments to claim a share of a tribal reserved water right. Although the Colville Tribes benefit from determinations made in the *Boldt* fishing litigation, any other reserved water rights claims made by the tribes have not been adjudicated or settled. The tribes have expressed their concern over sedimentation in Lake Roosevelt and the impact this has on tribal water use (Confederated Tribes of the Colville Reservation, 2000).

Significance for the Columbia River middle reach: The Colville Tribes are today focusing their economic development efforts on gaming and forestry operations, but the relatively large size of the reservation provides future agricultural opportunities. Any entitlement to reserved water rights for agricultural or other consumptive uses has not been adjudicated or settled; but if such rights are determined in the future, they would be senior to most downstream state law diversions and could diminish mainstem flows.

Warm Spring Indian Reservation (Oregon)

Pursuant to the 1855 Treaty with the Tribes of Middle Oregon, the Confederated Tribes of the Warm Springs Indians–comprised of the Wasco, Paiute, and Warm Springs bands—ceded 10,000,000 acres of aboriginal territory to the United States. Today, the Warm Springs Nation occupies a reservation of approximately 650,000 acres in north-central Oregon and is inhabited by 3,500 to 4,000 tribal members. The Deschutes River system, tributary to the Columbia, is the principal water source in the area. In an effort to avoid litigation, the Warm Springs Nation approached the State of Oregon in the early 1980s and offered to enter negotiations to determine, quantify, and settle its reserved water rights. After many years of negotiation, the final agreement was signed and executed on November 17, 1997. The agreement was submitted to the Deschutes County Circuit Court in 1999 for incorporation into the Deschutes River Decree, originally issued in 1928. In reaching the settlement, the parties had agreed *not* to use the "practicably irrigable acreage" standard that has been used in other water rights settlements and litigation. Instead, after studying 70 years of flow data from the Deschutes River, the parties thought that the region supplied enough water to satisfy all current and some future uses. The parties agreed that the amount of water resources used, consumed, and reserved as of September 26, 1996, was sufficient to satisfy their present and future water needs without subjecting other water users to a call by the tribes. The state subordinated its own instream flow right on the Deschutes River to the priorities of the tribal water rights.

Significance for the Columbia River middle reach: Although a reserved water right settlement has been reached for the Warm Springs Nation's claims on the Deschutes River, only the future non-Indian water development is constrained. The tribes are authorized to develop their water entitlement, and to the extent such development is consumptive, it will likely reduce flows in the Columbia River mainstem.

Flathead Indian Reservation (Montana)

The Confederated Salish and Kootenai tribes share the Flathead Reservation located near Flathead Lake in northwestern Montana. The tribes assert a variety of sovereignty and natural resource rights based on the 1855 Treaty of Hellgate. Land ownership arrangements on the Flathead Indian Reservation reflect a checkerboard-type pattern of Indian and non-Indian lands. The tribes and non-Indians living in the Flathead Valley have long contested the water supplied by the Flathead Indian Irrigation Project. The tribes have prevailed in many lawsuits concerning water, including a recent Montana Supreme Court decision preventing the state from issuing additional groundwater permits until a general stream adjudication for the basin is completed. This decision notwithstanding, an increasing number of unpermitted wells have been drilled. Although adjudication claims have been filed for water uses in the area, the adjudication has been stayed pending negotiations between the tribes and the Montana Reserved Water Rights Compact Commission. Those negotiations have barely commenced due in large part to heightened emotions on all sides. If negotiations are unsuccessful, many difficult and potentially volatile years of litigation are anticipated.

Significance for the Columbia River middle reach: The Flathead tribes have an ambitious economic development program, and their reservation is in one of the fastest-growing areas of Montana. Years, if not decades, will be required before existing water rights are clarified. Water use in the area will increase and is thereby likely to reduce flows to the middle and lower portions of the Columbia.

Nez Perce Tribe/Nez Perce Indian Reservation (Idaho)

The Nez Perce Indian Reservation in Idaho has the Clearwater River as its northern border. The reservation is also in the proximity of the Lochsa and Salmon rivers, as well as the Snake River itself. The tribe, and the United States on its behalf, has filed extensive claims in the Snake River basin. The claims are for sufficient instream flows to support salmon, as well as for water for irrigation and domestic uses. Instream flow claims have been filed in 1,134 drainages and extend virtually to all the water in the Snake, Salmon, and Clearwater basins (Shelton, 1997). The legal basis for the tribe's claim is its 1855 treaty,[6] in which the tribe reserved the exclusive right to fish all streams running through or bordering the reservation and a nonexclusive right to fish in "all usual and accustomed places."

In 1998 non-Indian water users filed a motion for summary judgment in the Snake River adjudication challenging the tribe's off-reservation instream flow water rights claims. In 1999 the trial court conducting the adjudication granted the motion for summary judgment and dismissed the tribe's and the U.S. instream flow claim, holding that no implied federal reserved instream flow right exists as a matter of law to support the tribe's fishery right [(Consolidated Subcase No. 03-10022 (Snake River Basin Adjudication Dist. Ct., Idaho, Nov. 10, 1999), appeal filed, Docket No. 26042 (Idaho Nov. 29, 1999)]. The tribe subsequently filed a collateral challenge to the ruling based on an alleged conflict of interest involving the judge, but the action was dismissed as moot after the judge resigned [*United States v. State*, 51 P.3d 1110 (Idaho 2002)]. The Idaho Supreme Court still has not determined the merits of the instream flow case, although the briefing was completed in February 2003. In the meantime, the major parties to the Snake River adjudication have been involved in mediating the Nez Perce claim. Reportedly, settlement discussions have focused on two major areas: (1) possible reconfiguration of the dam and reservoir system on the lower Snake River and the mainstem of the Columbia and (2) preservation of fish habitat in the Salmon and Clearwater basins (Shelton, 1997). On May 7, 2003, the Snake River adjudication presiding judge informed the mediating parties that he would order an end to the mediations and advance the remaining Nez Perce claims toward trial (Idaho Statesman, 2003).

Significance for the Columbia River middle reach: The Nez Perce have a senior treaty-based claim on some of the waters of the Snake River system. To the extent they are successful in having their instream flow claims recognized in the Snake River adjudication, Snake River and Clearwater flows at the Washing-

---

[6] Treaty with the Nez Perce, 12 Stat. 957 (June 11, 1855).

ton-Idaho border would likely stabilize or possibly increase. This would likely enhance water availability in the middle reach of the Columbia River.

Fort Hall Settlement (Idaho)

The Shoshone and Bannock tribes share the Fort Hall Reservation in southern Idaho. The Fort Hall Indian Reservation was established by an executive order in 1867. Initially intended to be 1,800,000 acres but later reduced to approximately 544,000 acres, the reservation is located along the Snake River near Pocatello. It is owned primarily by the tribes collectively (47 percent) and by individual Indian allottees (43 percent). In 1985 the state legislature directed the Idaho Department of Water Resources to commence a general stream adjudication in the Snake River basin. The legislature also passed a resolution, at the request of the Shoshone-Bannock tribes and the Idaho executive branch, authorizing negotiations to settle the tribes' water claims in the Snake River basin. The tribes and the state entered into a memorandum of understanding establishing a process for negotiating a settlement. The tribes obtained a special exemption from the U.S. Department of the Interior allowing them to pursue negotiations independent of the federal government. The United States and certain local water users were later included in the negotiations. In 1989 an agreement was reached that sought to protect the rights of water users established under state law. In late 1990 this agreement was ratified through congressional enactment [Fort Hall Indian Water Rights Act of 1990, P.L. 101-602, 104 Stat. 3059. See also Committee Report 101-831 to accompany H.R. No. 5308, 101st Cong., 2nd Sess. (1990)].

The settlement, involving a highly developed reach of the Snake River, sets the tribes' entitlement to water from the Snake River basin at 581,031 acre-feet per year. The water supply is comprised of a combination of natural flow, groundwater, and federal contract storage water. This entitlement satisfies all claims to water that the tribes may have had under the *Winters* doctrine. Indian rights in the Fort Hall Indian Irrigation Project were converted to *Winters* rights with a priority of 1867, the date the reservation was established.

Significance for the Columbia River middle reach: The Fort Hall Settlement is one of the few instances in which the *Winters* rights of an upstream Indian reservation have been determined.

FEDERAL RIGHTS AND OBLIGATIONS

Navigation

The federal government has plenary authority to regulate interstate commerce. Under the U.S. Constitution's interstate commerce clause, Congress may enact statutes regulating interstate commerce. The dormant interstate commerce power is also available to invalidate state statutes and other actions that impermissibly burden interstate commerce. One aspect of the interstate power is federal navigation power that enables the federal government to prevent obstructions that burden riverborne commerce on navigable waterways. The federal government's navigation authority prevents the construction of bridges or other structures that might impede navigation. It also prevents actions that deplete water so that navigation is no longer possible.

Significance for the Columbia River middle reach: Most of the mainstem Columbia River in Washington is navigable and is thus subject to the restraints imposed by the federal navigation authority. The federal government can always insist on a base flow in the river sufficient to allow actual navigation. The exercise of this authority trumps all state actions or diversions under state law that would interfere with this base flow requirement. Because the federal navigation power is constitutionally based, it may even limit federal statutes or federal agency actions that jeopardize navigation flows. Flows necessary for navigation on the mainstem of the Columbia and Lower Snake rivers may thus be the most legally secure water rights in the system.

Federal Reserved Water Rights (Non-Indian)

Hanford Reach National Monument

Non-Indian federal lands can also benefit from federal reserved water rights. In 2000, President Clinton signed an executive order creating the Hanford Reach National Monument, a 195,000-acre monument along the Columbia River in south-central Washington [Proclamation 7319, Establishment of the Hanford Reach National Monument (June 9, 2000)]. The site includes a 51-mile stretch of the Columbia River upstream of Richland. The monument designation was challenged in two separate lawsuits, but the U.S. Court of Appeals, District of Columbia, ruled in October 2002 that the designation had been proper under the 1906 Antiquities Act (16 U.S.C. § 431). The proclamation recognizes the importance of this reach of the river for fishery values. As discussed in a background paper accompanying the proclamation, the "[r]each contains islands, riffles, gravel bars, oxbow ponds, and backwater sloughs that support some of the most productive spawning areas in the Northwest, where approximately 80 percent of the upper Columbia Basin's fall Chinook salmon spawn. It also supports healthy runs of naturally-spawning sturgeon and other highly-valued fish species."[7] The proclamation specifically addresses water rights in the Columbia. It "reserves in the portion of the Columbia River within the boundaries of the monument, subject to valid existing rights and as of the date of the proclamation, sufficient water to fulfill the purposes for which the monument is established."[8] It also bans any new agricultural irrigation within the monument boundaries.[9]

Significance for the Columbia River middle reach: The Hanford Reach National Monument withdrawal creates a non-Indian federal reserved water right with a priority date of June 9, 2000. Among the purposes of the withdrawal is the reservation of water necessary to support spawning salmon and other fish species. This reserved right will prevent any new upstream consumptive diversions that would leave insufficient flows in the

---
[7] White House, Background Paper on the Hanford Reach National Monument at 2 (nd).
[8] *Id.* at 4.
[9] *Id.* at 5.

river to maintain the fishery protected by the reservation. As such, this reservation could be a significant constraint on new diversions upstream of the Hanford Reach.

Federal Regulatory Water Rights

Endangered Species Act

Mainstem water uses can also be limited by federal regulatory authority, sometimes referred to as "federal regulatory water rights." Because of the Columbia River's anadromous fishes, both the U.S. Fish and Wildlife Service and NOAA Fisheries have responsibilities for implementing the requirements of the Endangered Species Act in the basin. Between 1991 and 1992, Snake River salmon species were listed under the Act. Federal biological opinions issued in 1993 and 1994 were rejected by federal courts. A 1995 Biological Opinion established stronger protections, including increased flows and measures to improve indices (e.g., water quality and temperature) important for fishery resources. The Biological Opinion set a goal of adopting a revised opinion by the end of 1999. It also obligated the Corps of Engineers to prepare an environmental impact statement on breaching the four lowest dams on the Snake River (Ice Harbor, Lower Monumental, Little Goose, and Lower Granite; see Figure 3.1). The 1995 Biological Opinion was amended to incorporate additional protections as several other Columbia and Snake River runs have been declared threatened or endangered. Between 1995 and 1999, nine additional Columbia River basin fish species were listed under the Endangered Species Act (bringing the total number of listed populations to 12; see Table 1.1). In 2000 another Biological Opinion was issued for the Federal Columbia River Power System. In 2002 NOAA Fisheries concluded that federal agencies were successfully implementing 176 of 199 Reasonable and Prudent actions of the 2000 Biological Opinion requirements. The Biological Opinion was challenged in federal district court (Oregon) and in 2003 was found invalid. It was remanded back to the federal government for additional consulting and re-drafting. A revised Biological Opinion is expected to be issued later in 2004.

Significance for the Columbia River middle reach: The Endangered Species Act and the biological opinions produced under it are the principal federal regulatory constraints on federal agency actions affecting the Columbia River.

Federal Power Act

Since passage of the Federal Water Power Act in 1920, the Federal Power Commission and its successor, the Federal Energy Regulatory Commission (FERC), have been responsible for the licensing of hydroelectric power dams and facilities on navigable waterways. Typically, these licenses have been for 50-year periods. Two provisions of the act allow the FERC to impose license conditions protective of fish. Under section 10(j), FERC must impose conditions "based on recommendations received pursuant to the Fish and Wildlife Coordination Act from the National Marine Fisheries Service (today NOAA Fisheries), the United States Fish and Wildlife Service, and State fisheries and wildlife agencies" [16 U.S.C. § 803(j)]. Section 18 of the act also mandates that FERC "require the construction, maintenance, and operation by a licensee at its expense of . . . such fishways as may be prescribed by the Secretary of Interior or the Secretary of Commerce, as appropriate."

Significance for the Columbia River middle reach: The Idaho Power Company's "Hells Canyon Complex" of three dams (Hells Canyon, Oxbow, and Brownlee dams; see Figure 3.1) on the Snake River is a key Columbia River basin hydropower generating facility. This system is currently undergoing relicensing by the FERC, as its current license expires in 2005. An issue in the proceeding before FERC is how the dams should be operated or altered to protect salmon. It will thus be uncertain for several years how much water, and when, Idaho Power Company will have to release to protect instream values downstream of the dams. Instream flows below the Hells Canyon Complex will likely not be reduced during FERC proceedings.

STATE LAWS AND INSTITUTIONS

Near the beginning of the twentieth century, western states began to vest state administrative agencies with increasing amounts of authority to permit and manage the states' water resources. Many of these efforts were a reaction to courts that had allowed many western streams to become over appropriated. Many of the efforts resulted from the scientific management movement that sought to rationalize business and governmental processes. The efforts were also encouraged by the Progressive conservation movement that sought multiple uses of natural resources (Hays, 1959). With passage of the Reclamation Act in 1902, western states had an incentive to systematize their water rights records so they would be more competitive in securing federally supported reclamation projects.

Washington

Washington Department of Ecology

Washington was one of several states to reorganize governmental structure to better address the increased priority on environment issues during the 1970s. The Department of Ecology was established in 1970 with the goals to prevent pollution, clean up pollution, and support sustainable communities. Several smaller agencies were combined into a single department that encompasses a wide range of tasks, including among others water allocation, protection of water quality, and land use planning, jobs that are separated at the federal level and in many other states. The comprehensive holistic jurisdiction of the Washington State Department of Ecology allows the consideration of spill-over or second order effects of environmental decisions from one medium to another. For example, consequences of land use decisions may be traced to effects on air and water quality and water use within a single organization. With regard to funding for the agency's water resources program, budget year 2001 to 2003 included an appropriation that was lower than in the years 1993 to 1995. The program staff was reduced, including a reduction in water rights permit staff from 55 to 20. From

1997 to 2001, Department of Ecology enforcement staff was reduced from nine to one full-time equivalent.

Water Permit System

Washington State water law reflects a combination of the riparian water rights doctrine (generally used in the eastern United States) and the prior appropriation doctrine (used in different forms across the western United States). Although riparian rights initially framed the state's water laws, the state made a gradual transformation to the prior appropriation doctrine that culminated in 1917 with passage of a water code establishing permits as the exclusive way to obtain surface water rights. In 1945 the permitting system was expanded to include groundwater (with some exceptions). Although prior riparian rights were guaranteed in this legislation, the Washington Supreme Court later ruled that riparian rights not used by 1932 had been forfeited.[10]

A continuing problem in many western states has been the development of an adequate procedure for recognizing water rights established prior to or otherwise outside the state permitting system. In Washington these nonpermitted rights include rights established before the 1917 surface water code, groundwater rights established before the 1945 amendment, groundwater uses exempted from the 1945 act, riparian rights, and prescriptive rights (until this means of appropriation was abolished). In 1967 the state enacted the Water Right Claim Registration Act (later amended) allowing claimants to register these nonpermitted water rights. A timely and proper registration afforded the claimant with prima facie evidence of the quantity and priority of the claimed right. Failure to file a claim constituted a waiver and relinquishment of the water use. Since 1917, water adjudications have also been used to determine surface water rights, especially in basins where tribes and federal agencies assert reserved water rights claims. The largest adjudication involves the Yakima River basin, commenced in 1977 and now nearing completion.

According to the current permitting procedure prescribed by

---

[10] *Department of Ecology v. Abbott* (In re Deadman Creek Basin), 694 P.2d 1071 (Wash. 1985).

state law, the Department of Ecology cannot issue a water right unless four conditions are met:

- Water is available,
- The intended use is beneficial,
- The right will not impair existing water rights, and
- The public interest will not be harmed.

The importance of the public interest criteria is reinforced by Washington State statute [Washington Code of Regulations 90.54.020(3)(a)]:

> Perennial rivers and streams of the state shall be retained with base flows necessary to provide for preservation of wildlife, fish, scenic, aesthetic, and other environmental values, and navigation values. . . . Withdrawals of water which will conflict therewith shall be authorized in only those citations where it is clear that overriding considerations of public interest will be served.

The test for uses deemed in the public interest apparently considers the following:

- Consistency with the Department of Ecology, other state and federal natural resources management plans, and local land use and growth management plans. Consistency with applicable coordinated water system or utility plans;
- Effects on navigation, water quality, public health, and safety;
- The extent to which the proposal advances water conservation and efficient water use. Maximum net benefits to the state and region, including opportunity costs of foregone uses;
- The merits of the proposed allocation in comparison with alternative sources and methods of water development (including costs external to the applicant);
- The extent to which the use of water creates new burdens on the public agency for monitoring, regulation, oversight, and adjudication.

This public interest provision has been interpreted by the Washington Pollution Control Hearings Board, in cumulative effect situations, as follows:

When chronic water shortages have resulted in three water rights adjudications in a basin and reduced flows are depressing fish populations, even very minor irrigation applications may be validly denied. Though the effect of one small diversion may not be noticeable in isolation, the allowance of many such diversions would have a substantial impact. The potential for cumulative impacts may sustain a denial on public interest grounds. [(*Byers v. DOE*, PCHB No. 89-168 (1990); *Holubat v. DOE*, PCHB No. 90-36 (1990)].

Significance for the Columbia River middle reach: This interpretation of the public interest criteria is relevant to scenarios that posit additional diversions from the mainstem over the next 20 years. The rules emphasize the importance of cumulative effects and suggest that any individual diversion must be considered in the context of other likely calls on the river and environmental needs and changes. Once the permitting process is reopened, it may be expected that additional applications will be made from other sources in Washington. Also, if other upstream states anticipate the creation of downstream rights, this situation may provoke the filing of water rights applications in those states.

Instream Flow Protection Program

Washington's instream flow program originated with legislation passed in 1969.[11] Pursuant to this legislation, administrative rules were adopted by the Department of Ecology in 1980, and minimum instream flow values were established for the mainstem Columbia River upstream of Bonneville Dam.[12] The rules established minimum instantaneous flow requirements at five locations on the mainstem for 17 different time periods during the year. The rules also established minimum average weekly flows at five locations on the river for the same time periods.[13] In low water flow years the director of the Department of Ecology can reduce the minimum instantaneous and/or average weekly flows

---

[11] *See* WASH. REV. CODE §§ 90.22.010, -020 (2004).
[12] See WASH. ADM. CODE §§ 173-563-010 et seq. (2003).
[13] *Id.* § 173-563-040.

by up to 25 percent. However, outflow from Priest Rapids Dam can never be less than 36,000 cubic feet per second. Also, the Columbia River must provide at least 39,400,000 acre-feet per year at The Dalles.[14]

These instream flow rights have been recognized as appropriations with priority dates as of the effective dates of their establishment (1980 for the Columbia mainstem).[15] As such the instream flow rights are subordinate to "existing water rights, riparian, appropriative, or otherwise, existing on the effective date of this chapter, including existing rights relating to the operation of any navigation, hydroelectric, or water storage reservoir, or related facilities."[16] The instream flow rights are also subordinate to any water withdrawal at the request of the U.S. Bureau of Reclamation for the complete development of the Columbia Basin Project.[17] Approximately one-half of the Columbia Basin Project's authorized lands are not yet irrigated, and any water diverted for these new lands at the project would also be senior to mainstem, instream flow rights. The instream flow rights are also subordinate to any federal agency or tribal reserved water right established before 1980. Thus, this collection of various rights (existing pre-1980 rights, pre-1980 reserved water rights, and additional water withdrawn for the Columbia Basin Project) are essentially senior to the instream flow rights. They are also referred to as "uninterruptible water rights." Such rights include approximately 4,530,000 acre-feet of water rights based on state law.

The instream flow rules authorized the Department of Ecology to approve additional mainstem diversions, but they would be junior to the instream flow rights and subject to additional conditions imposed by the administrative rules.[18] For the first

---

[14] *Id.* § 173-563-050.
[15] *Hubbard v. Department of Ecology*, 936 P.2d 27 (App. 1997).
[16] WASH. ADM. CODE § 173-563-020(3).
[17] Shortly after passage of the National Reclamation Act in 1902, the Washington legislature authorized the United States to ask the state for withdrawal of water necessary for planned reclamation projects. This withdrawal was initially effective for one year but could be extended repeatedly if construction was under way. *See Id.* § 90.40.030. The legislature later allowed water to remain withdrawn for the ultimate development of the Columbia Basin Project, so long as the project was not abandoned. *Id.* § 90.40.100.
[18] The instream flow rules apply to public surface water and "any ground water the withdrawal of which is determined by the department of ecology to have a significant and direct impact on the surface waters of the main stem of the Columbia River" WASH. ADM. CODE § 173-563-020(1). Thus, certain post-1980 groundwater diversions are junior to, and can be administered to benefit, the instream flow rights.

4,500 cubic feet per second of water rights issued subsequent to the instream flow rights, these later rights are subject to priority administration if April to September flows at The Dalles are forecast to be 60,000,000 acre-feet or less *and* it is further predicted that minimum average weekly flows will not be met at one or more locations. Any water rights beyond the initial 4,500 cubic feet per second flow are subject to priority administration when the March 1 forecast of April to September runoff at The Dalles, Oregon (as published by the National Weather Service in *Water Supply Outlook for the Western United States*) is equal to or greater than 88,000,000 acre-feet and it is likely that minimum average weekly flows will not be met.[19] These post-1980 water rights, which are junior under some circumstances to the instream flows, are called "interruptible rights." "Interruptible rights" totaling 172,358 acre-feet have been issued (Gerry O'Keefe, Washington State Department of Ecology, personal communication, 2004).

In the spring of 1992 the Department of Ecology adopted emergency rules that withdrew unappropriated waters of the mainstems of the Columbia and Snake rivers from further appropriations. This moratorium was extended in 1994 in an effort to rebuild the anadromous fish population and to respond to Endangered Species Act listings. In the 1994 rule the moratorium was scheduled to expire in 1999 or when the Department of Ecology established an instream resources management program. However, the department has postponed new allocations pending the availability of additional information about the status of fish and expert opinion (including this report). In 1997 the Washington State legislature passed a law stating that the Department of Ecology could not use these minimum values to make decisions on future, new applications. However, approximately 300 water rights already issued out of the mainstem were subject to minimum flow requirements and could be interrupted as they were in the 2001 season. Because of the moratorium, it is difficult to estimate how large the demand for new permits on the Columbia River mainstem would become in Washington if the permit process was fully opened.

As part of Washington's Columbia River Initiative, there have been discussions regarding the permitting of uninterruptible

---

[19] *Id.* § 173-563-056.

water rights. The Department of Ecology is apparently considering the exchange of traditional, priority-administered appropriative water rights for "uninterruptible" water rights that would be exempt from normal rules of priority administration. Water law scholars generally agree that rigorous priority administration of water rights is rarely practiced in western states. In theory, and in some highly administered basins such as those in Colorado, priority-in-time administration is a hallmark of the prior appropriation doctrine. Holders of senior rights are entitled to the full amount of their appropriation before junior appropriators can divert water (so long as the "call" on the junior right would not be a futile effort, because of conveyance losses or other reasons, in actually delivering water to the senior user). Uninterruptible water rights would appear to jump to the front of the line in terms of state-administered water rights priorities.

The major advantage of uninterruptible rights is that they provide a greater certainty of water supply and encourage more efficient use and application of water. Apparently, these more efficient rights would be satisfied before legally senior water rights. The Department of Ecology is in a more informed position to assess the constitutionality of such as approach, but some senior water rights holders would likely claim a taking of the most valuable aspect of their water right—its priority. Also, some legal experts argue that conserved water is available to satisfy the unserved needs of junior users or is available for new appropriations. If the goal is to enhance instream flows, state law must ensure that conserved water is dedicated to the stream. Also, it is unclear how uninterruptible rights could be immunized from other uses and demands on the river unless base flows for salmon are diminished. Federal and state water quality and endangered species requirements may trump the exercise of uninterruptible rights. The State of Washington is not likely to be able to control upstream water development in Canada, on Indian reservations, or in other U.S. states. If upstream uses reduce instream flows in the Columbia River's middle reach, the guaranteed exercise of uninterruptible rights compounds the situation and potentially compromises the water necessary for healthy aquatic habitat and fisheries.

Significance for the Columbia River middle reach: One apparent legal basis for this initiative is a rules provision allowing the director of the Department of Ecology to allow "[f]uture authorizations for the use of water which would conflict with the provisions of this chapter [Columbia River main stem instream resources] . . . when it is clear that overriding considerations of the public interest will be served."[20] These new uninterruptible water rights would have seniority over the 1980 instream flow rights. They could not be curtailed to maintain minimum instantaneous flow or average weekly flow requirements of the instream flow rules. These new rights would be subordinate to other pre-1980 water rights. It is unclear how these new uninterruptible rights would be administered in relation to other mainstem rights established between 1980 and 2004.

In exchange for this jump in priority, the Department of Ecology proposes that the new uninterruptible rights be issued only on the condition that the water user employ state-of-the art water conservation technology. The Department of Ecology previously adopted a rule requiring that the authorized quantity of any new Columbia River mainstem water rights "accurately reflect the perfected usage consistent with up-to-date water conservation practices and water delivery system efficiencies."[21] The proposal would potentially increase the amount of water that could be diverted ahead of the instream flow protections. These rights would be in addition to the approximately 4,700,000 acre-feet of rights to water (apparently not including tribal reserved rights) that now may be exercised before the state's minimum flow requirements may be activated.

Oregon

Oregon has a more rigorous permitting procedure than most western states and also places more adjudicatory power in the state's Water Resources Department. Permits for new uses are submitted to the department. The department makes a preliminary review of the adequacy of the application and a proposed determination as to whether the application will be granted. If the proposed determination is protested, a contested case hearing is held before the department. Thereafter, a final agency deci-

---

[20] *Id.* § 173-563-080.
[21] *Id.* § 173-563-060.

sion is rendered. Oregon is conducting an adjudication of all pre-1909 surface water rights and all pre-1955 groundwater rights. The Oregon Water Resources Department reviews claims, holds administrative hearings, and files its proposed determinations with the state circuit court. The court reviews the findings, holds hearings on protests, and issues a decree officially upholding or modifying the department's conclusions. The state has completed 94 adjudications representing approximately 70 percent of the state. In 1975 the department commenced an adjudication of claims to surface water rights in the Klamath River basin.

Significance for the Columbia River middle reach: Because Oregon initiated its permitting program in 1909 and vested an administrative agency with the major role in adjudicating pre-1909 water rights, its inventory of water rights and the associated legal entitlement is better than most other western states. Even the reserved rights of the Warm Springs Reservation have been determined (see earlier discussion); however, the method of calculation (assigning to the tribes water in excess of 1996 non-Indian uses) leaves a large margin for future tribal development. Thus, while Oregon is in a rather good position in calculating existing rights and uses that affect the Columbia River, future development remains uncertain.

Idaho

In Idaho the Department of Water Resources (DWR) approves new permits and changes in existing water rights.[22] Since 1963, permits have been required for groundwater diversions. In 1971 this requirement was extended to surface water appropriations. Once water under a permit has been developed, the applicant submits proof of beneficial use and the Idaho DWR examines the use of water under the permit. If such use is deemed satisfactory, the DWR issues a license for the water right. The issuance of a water right license by the DWR is prima facie evidence of the existence of such a right and is binding on the state as to the right of such licensee to use the described amount of

---

[22] These provisions are set forth in Idaho Code tit. 42.

water. Once established pursuant to state permit and license procedures, a water right is real property under Idaho law and may be acquired by lease or purchase. Although instream flow may constitute a beneficial use in Idaho, only the state Water Resources Board may apply for and hold such a right. To address water rights not represented by licenses and permits, as well as federal reserved water rights, the Snake River Basin Adjudication (SRBA) is pending before state court. The SRBA encompasses most of the surface water in the state except for the Bear River basin and the state's panhandle region. Initiated in 1987, the SRBA has proceeded faster than most state adjudications but remains many years away from completion due in part to the large number of claims involved.

Significance for the Columbia River middle reach: Pending completion of the Snake River adjudication, existing water use entitlements are difficult to estimate. Fortunately, the reserved rights of the Fort Hall Reservation have been settled (see earlier discussion), and the potentially large claims of the Nez Perce are likely to be predominantly instream flow rights. Snake River flows are being affected by upper-basin groundwater uses; and because groundwater rights are not being adjudicated in the Snake River adjudication, the extent and effect of groundwater use will be difficult to measure and control. Idaho is a rapidly growing state with increasing amounts of economic activity, so it is expected that its future water needs will likewise increase.

Montana

Montana was one of the last western states to require permitting of all but the smallest water uses. Prior to 1973, water uses in Montana could be established under "use rights" (actual diversions of water) or optional state filing requirements. With passage of the Water Use Act in 1973, Montana adopted one of the most comprehensive permitting programs in the West. Except for small uses, permits issued by the state Department of Natural Resources and Conservation are required for surface water diversions and, unlike many other western states, for groundwater withdrawals. Although Montana is developing sound water rights records for post-1973 appropriations, pre-1973 water rights are a jumbled collection of water rights established under a hundred years of changing state legal requirements, compounded by the unquantified reserved rights appurtenant to many Indian reservations and federal land holdings. To rectify this problem, the state commenced (first in 1973 and then in 1979 in an expanded form), a statewide general stream adjudication of most pre-1973 surface and groundwater uses, including claims for federal reserved water rights. Although claims have been filed in all basins of the state, the adjudication pending before the Montana Water Court is proceeding slowly and relatively few final decrees have been entered for tributaries of the Columbia River system.

Under Montana law, adjudication of reserved water rights is stayed while the particular Indian reservation or federal agency engages in negotiations with the Montana Reserved Water Rights Compact Commission. Although several pioneering compacts have been reached throughout the state, negotiations with the Salish and Kootenai tribes of the Flathead Indian Reservation, one of the largest claimants on the Columbia River system, remain stalled. Those tribes assert a variety of instream and consumptive uses in a rapidly growing valley area in northwestern Montana.

Significance for the Columbia River middle reach: Both Montana's Clark Fork River and Flathead River systems provide large flows of water to the Columbia River mainstem. These tributaries are important water courses in the most rapidly growing region of Montana, the Stevensville-Missoula-Kalispell corridor. In projecting future water uses in upstream states, the Washington State Department of Ecology has provided no assumptions for Montana's future needs. Indeed, because of the incomplete general stream adjudication and inchoate nature of the claims associated with the Flathead Indian Reservation, water uses that might occur under existing water rights based on federal and state law and under future permits are difficult to predict. This uncertainty adds to the risk of additional permitting in the Columbia River middle reach.

SUMMARY

Applications for water withdrawal permits from the mainstem Columbia River, and from groundwater within 1 mile of the river, have been pending within the State of Washington for several years. Most of these applications are for the reach of the river between Grand Coulee and John Day dams. Permitting decisions must be balanced with the state's obligations to protect and enhance the environment, which includes salmon habitat. As this chapter has pointed out, Columbia River hydrology and salmon habitat along the river in Washington are also influenced by upstream water management activities and policies. The challenges involved in the State of Washington's permitting decisions are magnified by the fact that many upstream areas are likely to increase future water withdrawals, including British Columbia, Indian reservations, and the states of Idaho and Montana. New water permits in Washington may be subordinate, or "junior," to future water development in other upstream jurisdictions. As long as upstream development does not exceed Canada's ultimate entitlement and equitable state shares of interstate water, additional water use in Montana, Idaho, Oregon, and other basin states will be senior to new permits in Washington. In most cases, tribal reserved water rights will also have priority over these new state permits. With increases in water diversions—both in upstream areas and under new permits in the middle reach of the Columbia River—water available to salmon will diminish unless other regulatory programs, such as requirements of the Endangered Species Act, are triggered. These trends suggest that water resources managers and decision makers in the Columbia River basin would be well advised to explore ways to better manage existing water supplies, create more flexible management regimes, and better manage the numerous risks and uncertainties that attend salmon and water management. Basin entities, for example, could develop reversible management actions and approaches that are actively monitored and evaluated and that aim to meet new water demands in areas such as the middle reach of the Columbia River.

The next two chapters of this report examine the topics of better management of existing supplies, risks, and uncertainties. Chapter 6 reviews market-based approaches, such as water transfers, water banks, and conservation measures that are being used in many parts of the country, and Chapter 7 discusses strategies for better managing risks and uncertainties.

6

Better Management of Existing Water Supplies

Increasing demands for water in many areas across the United States, constraints on traditional engineering approaches to augmenting supplies, and concerns over environmental impacts of additional water withdrawals have prompted the search for nontraditional means for procuring new supplies of water to meet shortfalls during drought periods or to provide for more permanent uses. Market-based mechanisms have been implemented in many western states in an effort to lend greater flexibility to water allocation and to reallocate water to higher-value uses without increasing water diversions. This chapter examines water's economic dimensions as well as experiences with water transfers and other nonstructural measures that could be used to help augment supplies. These market-based measures have the potential to contribute to economic and human needs. Furthermore, because they focus on improved water use efficiencies, they do not require additional water withdrawals and can thus also contribute to viable salmon populations and a healthy Columbia River ecosystem.

THE ECONOMIC VALUE OF WATER

As discussed throughout this report, the waters of the Columbia River today sustain a wide variety of economic activities. Columbia River salmon populations have important commercial, recreational, and cultural values. The Federal Columbia River Power System provides an abundance of low-cost electricity that has been crucial to the region's economic growth. The Columbia River is important for irrigation, as it supports the Columbia Basin Project and hundreds of irrigation farms. The river provides water for municipal and industrial uses in the Tri-Cities of Washington. The Columbia River and its tributaries assimilate and carry away agricultural, industrial, and municipal waste. Given increasing demands for water from the Columbia River and its tributaries, it is important to understand how the value of water varies across each of these different types of water use.

Water resources in the western United States have traditionally been allocated across competing uses via legal or institutional means, not by markets. As noted in Chapter 5, water resources in the western U.S. are typically allocated by the prior appropriation doctrine, which tends to fix the allocation of water across a specific set of uses. In an attempt to add flexibility to the prior appropriation doctrine, traditional definitions of "beneficial use" are being reconsidered in many western U.S. states by specifying how water rights holders use water. This requires some understanding the economic value of water across a range of different uses. There is a rich literature on the value of water in a number of uses, including agricultural, industrial, municipal, recreational, and hydropower uses. Estimates of water value can be influenced by a variety of factors. These include measurement techniques employed, the nature of the data used in the assessment, and assumptions made in the estimation. Spatial and temporal aspects of water use also affect its value.

The economic definition of value is tied to the concept of *willingness to pay*. This concept holds that the value of an item is equal to what an individual is willing to pay for it (in monetary terms) or in terms of what the individual would give up to obtain the item. This concept of willingness to pay is also related to the notion of "demand" and is related to the relationship between the demand for a good and its price. Specifically, a price-demand relationship can be viewed as an expression of marginal willingness to pay for the item (the term "marginal" refers to the value of the next or incremental unit demanded). This marginal willingness to pay usually declines with units consumed. In addition to the direct measurement of marginal willingness to pay for water, the concept of alternative cost can be used to assign values to water in various uses. With this concept, the value of water is defined as the cost of the least expensive alternative to water (Gibbons, 1986). The following values for various use categories are derived primarily from a review of the literature (Gibbons, 1986), in which values from several studies in each water use category were synthesized. Values listed in this section are expressed in 1999 U.S. dollars unless indicated otherwise.

Agriculture

In discussions regarding the value of water to agriculture, it is important to note the assumptions that underlie the procedures used to assign a value to water, as these assumptions influence the derived water values. Historically, as western water was allocated primarily in accordance with the doctrine of prior appropriation, and not by market mechanisms, there were thus no market prices from which to determine its "value." As a result, initial efforts at valuing agricultural water usually relied on techniques that imputed or inferred a value by comparing all expenses associated with producing a crop with the revenues received from sale of the crop. The residual value (the difference between revenues and assigned costs) was assigned to the unpaid input—in this case water. However, the residual value reported in some studies may also include other values, such as a return to the farmer's management as well as to land. It is thus important to claim only the residual due to water in assigning a value to water. The values reported in Gibbons (1986) appear to be for those associated only with water, as are additional references cited below. Another factor that affects water values is whether the value is assigned to water diverted (applied) to the field or assigned to water actually consumed (water "consumed" refers to the amount of evapotranspiration, or ET). Since diversions always exceed evapotranspiration, water values calculated using diversions will be lower than those based on ET. The values in this section are assumed to be based on diversions.

Water's value as an input in the agricultural production process depends primarily on the value of the crop being produced. Thus, a farmer's demand for irrigation water is a derived demand that depends on the demand for the crop being sold. The effect of crop value on water value is confirmed in numerous studies that have shown that the marginal value of water is higher for high-value crops than for low-value crops. For example, several studies conducted at different locations in the western United States have presented estimates for the marginal value of water for grain sorghum (a low-value crop) in the range of roughly $3 to $40 per acre-foot (1999 dollars), while estimates of the marginal value of water in the production of fresh vegetables (high-value crops) often exceed several hundred dollars per acre-foot.

Although many studies provide crop-specific marginal values of water, other studies estimate marginal water values based on current proportion of acreage dedicated to each crop type at a given location. For example, in a study of irrigated farmland in Oregon's John Day River basin, the U.S. Bureau of Reclamation estimated the value of water for the production of a mix of crops including pasture, alfalfa, and wheat to be in the range of $20 to $48 per acre-foot (Adams, 1999). At a different location in Oregon, Adams and Cho (1998) reported values for four regions of the Klamath Irrigation Project in southern Oregon and northern California. In a region of the project dominated by low-value crops, the marginal value of water across the crops in the region was estimated at $42 per acre-foot; in another region dominated by high-value crops, the marginal value of water across the set of higher-valued crops was estimated at $80 per acre-foot (ibid.).

The marginal value of water depends not only on the value of the crop to which it is applied but also on the quantity of water used by the crop and the nature of crop yield and water response relationships. Although there is debate over precise relationships, as more water is applied, the effect on yield generally begins to decline. Also, as efficiency (the proportion of water applied to the crop actually used by the plants) increases, one expects that the value of the water (or willingness to pay for water) will increase. Empirical evidence of this effect is found in a study in which marginal values for a representative Columbia River basin crop mixture were inferred to be $46 per acre-foot when water was tightly restricted but were only a few dollars per acre-foot when water available to the crop was not restricted (Bernardo and Wittlesey, 1989).

The range of the value of water in agricultural applications in the westernUnited States generally varies from values as low as $3 per acre-foot for low-value crops under conditions of adequate water supplies (no water stress) to values in excess of $200 per acre-foot for high-value crops. Median values for most mixed cropping systems in the Pacific Northwest suggest that the agricultural value is in the $40 to $80 per acre-foot range. For example, in a recent study of the economic impact of the scenarios defined in the Washington Department of Ecology's Columbia River Initiative (CRI), Huppert et al. (2004) estimated a value for additional agricultural water of $32 to $101 per acre-foot. The authors assumed that any new allocation of Columbia

River water under the CRI will be used on high-value crops (primarily orchard crops). It should be noted that farmers will be less likely to plant high-value irrigation crops with "interruptible" water rights, given the risk associated with loss of investment in drought years. This pattern of risk aversion is observed throughout other regions of the West, where farmers with junior water rights tend to favor lower-value crops. This is especially the case when water supplies vary substantially from year to year, as junior appropriators may have their allocations cut off under conditions of limited water flows and supply.

These values are estimated values, based on various economic assessment methods. These values are supported, however, by recent real-world experiences with water bank transactions in the western United States. For example, within the California Water Bank, created in 1998 and 1999 to address water shortages due to drought, equilibrium prices for water transfers between irrigators were approximately $75 per acre-foot. The actual value to irrigators may be slightly lower than this price given that the sales price includes a "tax" to provide water for environmental uses, primarily in the Sacramento-San Joaquin Delta. In the Klamath River basin in southern Oregon and northern California, a pilot water bank program created for the 2003 irrigation season also established a price of approximately $73 per acre-foot for the purchase of water from irrigators for environmental uses (this is the value averaged across both high- and low-value crops). During a drought in 2001, a temporary water bank was created within the Bureau of Reclamation's Yakima Project in south-central Washington. Substantial quantities of water were transferred from irrigation districts with more senior rights and low- value crop mixes, to districts with junior rights and higher value crops. For example, the Roza Irrigation District, which is dominated by high-value perennial crops such as tree fruits, purchased over 16,000 acre-feet of water at a season average price of approximately $120 per acre-foot (Northwest Economic Associates, 2004). In summary, patterns of water values observed in actual transactions in water banks in California, Oregon, and Washington are generally consistent with those that would be suggested by economic theory and data (higher water values for higher-value crop mixes), and they provide general corroboration of the estimated values (cited previously) of water found in the economics literature.

This information from economic assessments and actual transactions establishes a general set of values for irrigation water. Recently, however, the U.S. Supreme Court approved values for water in irrigation at substantially higher levels than those generally found in the economics literature or in market transactions. Specifically, in the case of *Kansas v. Colorado* (533 U.S. 1 (2001)), which concerned a dispute over Arkansas River water used for agricultural irrigation, the U.S. Supreme Court accepted values of approximately $125 per acre-foot (1999 dollars) for water used on a mix of wheat, corn, grain sorghum, and hay (low- to medium-value crops). These values were estimated by the State of Kansas as part of the damage assessment phase of the trial, and were accepted by the Special Master.[1] The values were accepted initially by the Supreme Court's special master in the case and ultimately approved by the court as part of the damage assessment. The implications of the values accepted by the Supreme Court for agricultural water may be significant, particularly in litigation concerning reductions in irrigation water deliveries to agriculture arising from state or federal policies or actions.

Municipal

The marginal value of water for residential purposes depends on the end use and the level of current consumption; marginal value is typically less for outdoor consumption (e.g., lawns) than for indoor consumption and typically declines as more water is consumed. In an early survey of water value estimates, Gray and Young (1983) found that published household valuations of water ranged from $63 per acre-foot for lawn watering to $403 per acre-foot for indoor water use. Gibbons (1986), on the other hand, synthesized three water demand studies and used the estimated demand equations to calculate marginal willingness-to-pay estimates. Estimates in that study ranged from $34 to $56 per acre-foot for summer consumption (primarily outdoor uses) and from $50 to $212 per acre-foot for winter consumption (pri-

---

[1] The values were developed for the State of Kansas and accepted by the special master and are for direct effects only; that is, they are representative of the effects on farmers' incomes only and thus do not include secondary effects on the local economy that may arise from reductions in water allocated to farmers.

marily indoor uses).

In a water transfer agreement negotiated in California in 2002 between the Imperial Irrigation District (IID) and the San Diego County Water Authority (SDCWA), water ultimately intended for municipal uses was valued at a minimum of $230 per acre-foot. This price equals the cost of conserving the water plus an incentive to encourage program participation by Imperial Valley farmers. The water's price reflects considerable effort by the IID and SDCWA to assess the cost of on-farm conservation measures, including systems to capture and reuse water, as well as line earthen irrigation canals with concrete. The actual cost of water delivered to residential users in San Diego will be substantially higher than $230 per acre-foot. The Imperial Irrigation District-San Diego agricultural-urban water transfer is a good example of how conserved water can be transferred to a use of greater economic value and how water supplies might be augmented in order to sustain economic growth without increasing withdrawals from surface or groundwater supplies. As with all water transfers, "third-party" effects should also be considered in the interests of equity.

Industrial

Industries utilize water for cooling, processing of products (e.g., washing materials, conveying inputs, input in the end product), in-plant sanitation, and other purposes. Since water costs constitute a very small portion of industrial costs, industrial demand for water is expected to be quite inelastic (i.e., there is little change in demand with changes in price). The amount of water used by industry is influenced by raw material quality, relative price of inputs, output mix, and government regulations. The cost of water to industry includes intake costs, treatment of water for recirculation, and waste treatment of effluent. When the price of water rises, firms typically reduce intake and increase treatment of water for reuse. Thus, the marginal value of water for industry is often estimated by the alternative cost of internal recirculation of water (Gibbons, 1986). Alternative costs of recirculation depend on the use to which the water is applied and on current processes. As water efficiency of the current technology increases, the marginal value of water also generally increases. Process recycling costs also vary widely by industry and current processing technology. One study of a textile finishing plant (Kollar et al., 1976) estimated marginal costs of $269 per acre-foot to increase the percentage of process water recycled from 48 to 76 percent, while another study investigating a meat-packing plant (Kane and Osantowski, 1981) with an extensive water reuse system estimated marginal costs of recycling process water at $660 to $939 per acre-foot. As water users become more efficient, the marginal value of water in industry rises.

Hydropower Generation

The Columbia River and its tributaries power one of the world's largest hydroelectric systems. In 1998 for example, the system produced an average of 12,000 megawatts of electricity, enough to supply a city 10 times the size of Seattle. Reductions in streamflow have important implications for the value of water in hydropower production. It is interesting to note that the marginal value of water for hydropower depends on where the water is in the Columbia River system; the higher the elevation of the water, the higher its marginal value, as water at a higher elevation in the system will generally pass through more generation facilities.

One study estimated marginal values of water in the Columbia system at various points along the river (Hastay, 1971). The study estimated water values for energy generation based on the alternative cost of requiring more thermal power generation to replace reduced hydropower generation. Marginal values of water were estimated at $4.50 per acre-foot at the downriver location of McNary Pool and a marginal value of roughly $20 per acre-foot at Upper Salmon (Butcher et al., 1972). A study by McCarl and Ross (1985) estimated the hydroelectric value of Columbia River water by calculating how much electricity costs would rise due to additional water being diverted for irrigation. The alternative cost of requiring more thermal generation to replace the decreased hydroelectric generation was found to range between $14 and $76 per acre-foot of additional irrigation diversions. The higher values corresponded to the value of water diversion farther upriver, while the lower values were based on water located in the middle reach of the Columbia River. In a study of Canadian hydropower, Gillen and Wen (2000) defined water's economic value differently. They defined the economic rent (marginal value) per kilowatt-hour of electricity from hydropower to be the difference between the competitive market price of electricity and average costs to produce electricity. Using long-run electricity supply contracts to estimate the competitive price and power utility financial records to estimate costs, the authors estimated cost savings arising from hydropower produced by Ontario Hydro at 3.4 cents per kilowatt-hour relative to the price per kilowatt-hour from other sources. Applying this measurement of value to the Columbia River system, the loss of its hydropower would imply roughly a doubling of electricity costs to the region if alternative means of power generation (i.e., fossil fuel) were required.

Recreation

The value of water for recreation is based on the value of recreational activities taking place both on the water (e.g., boating, fishing, windsurfing) and adjacent to water (e.g., picnicking, camping). Studies of the marginal value of water for recreation indicate that estimates of water values differ substantially, depending on recreational activity, magnitude of streamflow, and quality of the water. In a study of reservoirs in Colorado, the average recreational benefit of water retained in the reservoirs was estimated at $72 per acre-foot for each additional day retained (Walsh et al., 1980). A study by Ward estimated the value of water for angling and white-water boating in the Rio Chama River in New Mexico at $46 per acre-foot. A study of recreation in Colorado estimated marginal values of water at alternative streamflow levels. Marginal values of $41 per acre-foot for fishing, $10.25 for kayaking, and $7.70 for rafting were estimated along one stream stretch (Walsh et al., 1980). The dependence of water values on the levels of stream flow is evident in a study by Amirfathi et al. (1974) of angler benefits on northern Utah's Blacksmith Fork River. Marginal benefits were zero when flow was reduced by 50 percent but increased to $130 per acre-foot when flows were to 20 to 25 percent of peak levels. Studies often estimate the value of recreational fishing in terms of value per visitor-day. The more fish of a given species present in a body of water, the more anglers it can support and the higher the total value of water. The value of water for fishing per angler also depends on the number of other anglers present. The value of the fishing experience will likely be lower when use of the river by other anglers is greater (Lin et al., 1996).

Navigation

The Columbia and Snake rivers can be navigated as far up-river as Richland, Washington, and Lewiston, Idaho, respectively, the latter of which is 465 miles from the Pacific Ocean. Combined with barge traffic from the Willamette River, these stretches carried approximately 38 million metric tons of cargo into Portland in 2000, which represents approximately 5 percent of the Portland metro tonnage from all sources of transportation (Bingham, 2002). The majority of this cargo was grain: in the period 1990 to 1998, between 35 and 50 percent of all grain receipts at Columbia River terminals were shipped by barge on the Columbia River system, with the remaining portion shipped primarily by rail (Casavant, 2000). Waterway transportation can be advantageous because of the relatively low cost of transporting bulky, low-value commodities such as grain.

Short-run estimates of water value for navigation typically utilize the alternative cost method; the value of water used to support navigation is equal to the savings of using water-based transportation over railroad transportation, minus the costs of operation and maintenance of the waterway. Long-run estimates include the costs of construction of the waterway (it is assumed that railroad rates reflect all fixed and variable costs, while barge rates reflect only private costs and not waterway costs, since user fees are uncommon). The marginal value of water is either equal to zero (at all levels except the level at which the water flow is reduced such that navigation is no longer possible) or equal to the entire economic value of navigation (the level at which navigation is made possible). Therefore, average values (as opposed to marginal values) are typically used to estimate the value of water for navigation.

Ecosystem Goods and Services

The Columbia River provides an abundance of goods and services that include goods like food and fiber (salmon and other aquatic species), drinking water, services such as waste assimilation, and broader values such as biodiversity and aesthetic pleasure. These goods and services sustain important economic activities, such as commercial and sport fisheries.

Nonmarket Values

In addition to the direct economic value derived from the use of water and other ecosystem goods and services, there is a demand for values from the river that are not exchanged in markets. For example, the Columbia River system provides habitat for many valued fish species and also sustains populations of waterfowl, aquatic mammals, and other wildlife. Fish and wildlife provide nonconsumptive values to photographers, hikers, and others who enjoy outdoor recreation. People may also value the existence of salmon and other species in the Columbia system even when they do not directly observe or "use" them (so-called existence or nonuse values). Although it is more difficult to estimate existence values than values associated with direct use, numerous studies have shown that people express a positive willingness to pay for preserving ecosystems and the species within them. For example, in a study of passive-use values for coho salmon in the Columbia River, Olsen, et al. (1991) estimated passive-use value of $21.80 for each adult coho male that reached its natal stream. Huppert et al. (2003) reported a range of existence values for salmon in the Pacific Northwest of $66 to $268 per acre-foot of water. Although existence values appear to be very site and context specific, studies of existence values for other species and ecosystems suggest that the value of the waters in the Columbia River system in providing habitat for diverse species is high and of importance when making public policy decisions concerning the basin's water resources. Ideally, when comparing the efficiency of alternative water allocations, policy makers should obtain estimates of the sum of all use and nonuse values to determine the "total economic value" of a particular water allocation.

Cultural Values and Other Public Goods

Another category of passive-use values includes what are sometimes referred to as symbolic or lifestyle values, which account for the economic impacts imposed on areas of origin in water transfers. These values can relate to traditional means of livelihood and to the maintenance of ways of life and social cohesion (Brown and Ingram, 1987; Howe and Ingram, 2002). Farm families, often going back several generations, place a high value on a ranch or farm lifestyle and sometimes "stick it out" even when the economic activity becomes unprofitable (Weber, 1990). This category of values also includes symbolic values that may be placed on an undiminished river or stream. In the context of the Columbia River basin, salmon have particularly important cultural and symbolic values. Although not quantifiable in economic terms, these types of values often enjoy strong political support.

The cultural values inherent in the Columbia River system are part of a larger category of services and values described as "public goods," which have three broad features: (1) one person consuming them does not prevent another person from consuming them ("nonrival"); (2) if one person can consume them, it is impossible to prevent another person from consuming them ("nonexcludable"); and (3) people cannot choose to not consume them even if they want to ("nonrejectable"). Public goods are not normally provided by the private sector because there is no way to charge consumers for the provision of such goods (due to their nonexcludable nature). As a result, public goods might not be provided at all if left to market forces. Examples of public goods include flood control, clean air, and national defense. Many ecological goods and services from the Columbia River, such as the benefits of aquatic habitat and clean water, have features of public goods. Some public goods are provided in part by the government and are paid for through taxation. An example in the Columbia River basin is the habitat restoration programs funded through annual expenditures of the Bonneville Power Administration. Such public expenditures (to which hundreds of millions of dollars are devoted) illustrate that society greatly values these types of services, which suggests that consideration of public goods should be part of any debates and decisions regarding appropriate uses of Columbia River water and associated resources.

Summary

There are substantial differences in water values across different categories of water use in the western United States and across the Columbia River basin. Table 6-1 summarizes these values by use category. The significance of the differences across uses is that there is a great potential to promote economic growth and increase overall social benefits by transferring increments of water between uses (from low- to higher-value uses). The actual benefits from transfers will depend on the quantities of water transferred or diverted and the costs of such transfers. The following section examines means by which market-based mechanisms might help effect those transfers, some limitations of market-based measures, and some examples of their application across the West.

WATER MARKETS AND WATER BANKS

States across the western U.S. generally began using water markets in the 1970s as a means to address some of the inflexibilities inherent in the doctrine of prior appropriation (NRC,

TABLE 6-1 Marginal Values of Water in the Columbia River Basin

Use	Value Range per Acre-foot (1999 U.S.$[a])
Agriculture	$3-$200
Municipal	$34-$403
Industrial	$10-$1248
Hydropower	$4-$62
Recreation	$7.70-$130
Navigation	$5.60
Waste assimilation	$0.20-0.28
Passive uses	Not available on an acre-foot basis

[a]Converted into 1999 U.S.$ using Consumer Price Index annual figures.

1992). Water markets and water banks were developed as a means to reallocate water from lower-value uses to higher-value uses, or to environmental uses. The term *water market* refers to the temporary or permanent transfer of a water right or a contract entitlement for the use of water. The term *water bank* generally infers two types of arrangements (Miller, 2000). One can be labeled a groundwater storage bank. For example, in California's Kern County, the Kern County Water Bank provides for the purchase and underground "storage" of water in wet years, with that water then available in dry years for sale to the state. The process of recharging groundwater through wells into an aquifer for later use is also referred to as aquifer storage and recovery and its applications are being explored in other parts of the country, as well, such as the Florida Everglades (NRC, 2001b). *Water bank* also refers to a formal mechanism created to facilitate voluntary exchanges of the use of water under existing rights.

The rationale behind the water market concept is that willing buyers and sellers should be allowed to engage in mutually beneficial transfers of water. For example, an individual who lacks water rights or holds junior rights may be willing to pay more for water than an individual with superior water rights can realize by using the water. In such a case, both parties would gain from a trade or transfer of water and society would have realized greater value from the water through this transfer. To facilitate the creation of water markets, western states have changed laws and rules associated with the doctrine of prior appropriation to allow a water right to be separated from the land to which it was originally applied. In such cases the right is redefined as a particular flow or volume of water instead of a diversion at a particular location. Thus, under a water market or bank, a downstream user can purchase or lease water from an upstream user. The magnitude of the gain from such a transaction is determined by the seller's increase in returns (over the value of the water generated from use on site) plus the additional increase in income or averted loss realized by the downstream purchaser. It is assumed that trades will not occur unless they are of mutual benefit to buyer and seller. The existence of a market also allows other prospective water users, such as parties who currently hold water rights, to obtain water previously unavailable to them. For example, conservation groups or fishery agencies may purchase water to maintain instream flows. In Washington State the Department of Ecology has created a program to acquire water for instream uses. In addition, the Washington Water Trust (a private organization) has been acquiring water for instream flow purposes. In some western states (e.g., Colorado, Arizona), municipalities purchase agricultural water rights through water markets to meet rising water demands driven by human population growth.

Although most water markets are intrastate, an interstate water market was recently authorized by the federal government to allocate waters of the Colorado River. In October, 1999, then Secretary of the Interior Babbitt issued rules authorizing trading of Colorado River water among Arizona, California, and Nevada of Colorado River water. The plan calls for Arizona to act as the "bank," building on the Arizona Water Bank (created in 1996) experience. The significance of this rule is that it sets a precedent for states to develop joint water markets. The advantage of broadening the scope of a water market is to create more opportunities for trade and the prospects for realizing greater social benefits. Such an interstate water bank seems well suited to improving water use efficiencies in the Columbia River basin.

Limits of Markets

Water banks hold the potential to increase social benefits associated with water uses, but they contain some limitations, and their successful implementation poses challenges. A key consideration in water transfers is the notion of *third-party effects*. Most water uses do not consume all water that is diverted. Some portion of unused water moves back to the stream through surface flows or it percolates to groundwater, where it becomes available for other users. Water transfers may disturb this pattern of return flows and have effects on individuals or groups ("third parties") outside a market-based transfer. These third-party effects can be remedied by adjusting and reducing the amounts of the water right available for transfer. There are situations, however, when third-party effects are difficult to quantify and monitor. Further, parties may be so unequal in resources and bargaining power that some sales are at least potentially coercive. In a study of water sales in eight western states, it was found that 90 percent of the water exchanged through markets went to municipal, federal, and state agencies and that 96 percent of transactions involved these relatively large and powerful participants (Brookshire et al., 2003). Farmers looking to sell water typically are not equal in resources, in negotiating experience, or in bargaining power, especially when there is only one large possible buyer. The city of Tucson, for example, was able to purchase most of the groundwater rights in nearby Avra Valley, transferring water use from agricultural to urban uses at modest prices that left many farmers with bitter feelings toward the city. Avra Valley farmers believed that as long as the city was buying up water rights, the rural economy had no real future, and individual farmers feared that if they did not sell, their neighbors would undercut their price. In addition to these types of potential drawbacks of water transfers, there can be significant transaction costs associated with locating willing buyers and sellers, legal services, and hydrological studies (Miller, 2000).

Damages to localized rural economies that result from large-scale water transfers to urban areas are real and can be significant. For example, seed, fertilizer, and implement sellers, as well as retailers, suffer when large portions of their customer base disappear. Government services are adversely affected in rural counties when tax rolls decline. Damages can be partially mitigated by "area of origin" protections that provide for transfer payments made from urban to rural counties involved in water transfers. There are values associated with water that are often poorly reflected in market transactions. Water in the western U.S. has long been associated with opportunity: areas with ample amounts of cheap water available for development have a future, while those that do not have water face a more uncertain future. Rural areas of origin in water transfers often perceive that they have compromised their future. Cultural values associated with water are among public values not likely to be protected in private water markets. Some commentators mourn the changes in the interior West where agribusinesses have replaced family farms and the vast majority of the population now lives in cities (Little, 2003), and there are fears that water markets will facilitate this transformation.

Infrastructure for Water Markets

Water markets are only as good as the governmental authorities that regulate them. But simply facilitating the creation of markets will not allow governments to get out of the business of managing water. On the contrary, water markets must be managed just as intensively as water permitting programs, and additional skills and infrastructure are required for their effective execution. In fact, water markets must be based on secure property rights or permits, and permitting authorities are implicated in the task of quantifying transferable water made uncertain by changes in uses affecting points of diversion, return flows, and instream flows. For water markets to operate effectively, information should be freely available and transactions within water markets should be transparent. Governments have important roles to play in ensuring this transparency by facilitating a free flow of information to prospective sellers and buyers. At a minimum this involves establishing real-time information databases or electronic bulletin boards that reflect ongoing market transactions. Government also has a role as a monitor and referee in lease and option arrangements. Additional resources may be necessary in the Columbia River basin to help state agencies perform these duties if such markets are pursued.

Governments as participants in water markets need skills and resources that go beyond those required in regulating markets. Where state governments enter markets to buy or lease water for fisheries restoration or other instream uses, state agencies must build skills in buying and selling water, which involve financial as well as ecological risks. For example, fishery managers have faced a steep learning curve in operating the Environmental Water Account through which water to protect fish habitat is acquired in the California CALFED Bay-Delta Program, which is a joint state-federal partnership for managing the San Francisco Bay-Sacramento River-San Joaquin River delta. Fishery managers have been accused of paying too much for some water as well as hoarding water as a hedge against uncertainties that might occur later in the year rather than releasing it to save fish endangered early in the season. The human resources requirements for effectively supporting such activities include staff with backgrounds in business, economics, and marketing—skills and expertise typically not widely found in most natural resources agencies.

Applications

Applications of water markets and water banks are increasing across the western U.S. Many of the 17 western states presently allow water to be sold or leased. The use of water banks, in particular, especially increased during the 1990s. Since water banks typically involve the temporary transfer (lease) of a water right, they can be particularly useful during drought periods. Water banks may also ameliorate some of the negative effects of a permanent transfer of a water right. Farmers and rural communities may thus be more receptive to the water bank concept than to sales of water rights (Keenan et al., 1999).

Water banks hold promise for water problems such as those that recently occurred in the Klamath River basin. A 2002 U.S. Bureau of Reclamation Biological Assessment and related operating plans for the Klamath Project call for the creation of a water bank of up to 100,000 acre-feet of water per year. This water would come from groundwater sources and from surface water obtained by idling lands within and outside the project. Funding for purchases of such water would be provided by the Bureau of Reclamation. The "banked" water would be used for environmental purposes, primarily to maintain water levels in upper Klamath Lake and streamflows on the lower Klamath River. The State of Oregon, however, has not finished the adjudication process for water rights in the Klamath River basin. In the short term, water banking will need to rely mostly on water sales among Klamath Project farmers in the California portion of the basin who may have water available for transfer, such as from wells.

The State of California has pioneered several market-based programs and agreements aimed at shifting water among users, with many good results. For example, in response to a pronounced drought that started in the late 1980s, the state established several emergency water banks that were viewed by many as successful at redressing imbalances between availability and demands during shortages (Miller, 2000). For example, one study estimated the net benefits of the 1991 water bank at $91 million, with net benefits of $32 million to the agricultural sector

(Howitt et al., 1992). Water for the 1991 water bank came from three sources: fallowing, groundwater, and surface storage. This and other water banks in California are managed by the State Department of Water Resources.

In December 2002 the Imperial Irrigation District (IID) and the San Diego County Water Authority (SDCWA) approved an agreement for the long-term transfer of conserved water from the Imperial Valley to the San Diego region. This agreement is a principal component of the Quantitative Settlement Agreement, California's plan to abide by its Colorado River water allocation (California has long exceeded its legal allocation by about 20 percent). Under this agreement the IID and its agricultural customers would conserve water and sell it to the SDCWA for at least 45 years. Deliveries in the first year of the contract would total 20,000 acre-feet and would increase annually in 20,000 acre-foot increments until they reach a maximum of 200,000 acre-feet. In the event of water shortages in the Colorado River, the IID and the SDCWA would share shortages proportionately. The price of the transferred water between IID and the SDCWA is currently set at $248 per acre-foot. This price equals the cost of conserving the water plus an incentive to encourage participation by Imperial Valley farmers. The water's price reflects considerable effort by the IID and the SDCWA to assess the cost of on-farm conservation measures, including systems to capture and reuse water and line earthen irrigation canals with concrete. Specifically, price is calculated in the contract by a formula that indexes the water's price to the Metropolitan Water District of Southern California's water rate minus the SDCWA's cost to transfer the water to San Diego County. A discount is applied to the price that begins at 25 percent in year one and declines gradually over 17 years to stabilize at 5 percent for the remainder of the contract. Under this formula, water price is comparable to that of other supplies available to the SDCWA (*www.iid.com/water/transfer.html*, accessed Janauary 11, 2004).

Four state- and federally funded water transfer programs exist or are being developed in California to facilitate water transfers to the environment. The projects are the Environmental Water Account (EWA), the Environmental Water Program (EWP), the Water Acquisition Program (WAP), and the Drought Water Bank. The EWA and EWP are part of the CALFED Bay-Delta Program, which is a cooperative effort that, among other goals, is addressing environmental problems within the aquatic ecosystems in California's Bay and Delta region. The WAP formed under the authority of the Central Valley Project Improvement Act (CVPI), is a U.S. Department of the Interior joint program of the Bureau of Reclamation and the U.S. Fish and Wildlife Service. The WAP acquires water for protecting, restoring, and enhancing fish and wildlife populations to meet the goals of the CVPIA. Table 6-2 shows the water acquired between 1993 and 2001 by the WAP. As indicated, substantial amounts of water have been transferred for fish and wildlife purposes, at a wide range of prices.

The EWA was established to make additional water available at critical times during the life cycle of various endangered and threatened species, while not adding additional costs or uncertainty to urban or agricultural users. The Environmental Water Account has a portfolio of variable and fixed water assets. It acquires water from willing sellers and banks borrowing and transferring water from one location to another. In the three years it has been operating, it has helped provide security to users while also allowing fishery managers additional water supplies at critical times. The main criticism has been that EWA managers may have paid too much for water at certain times, although it is reasonable to expect that agencies with limited experience in water markets may make some mistakes as they gain experiences with these processes.

The Columbia River basin has had some experiences with market-based water transfer mechanisms. The State of Idaho, for example, has implemented a water banking scheme. The Idaho scheme differs somewhat from the water banking system managed by California's Department of Water Resources. The Idaho water banking program aims to help irrigation districts earn a modest return from sales of surplus water in wet years and to keep water in irrigated agriculture during drought years (Miller, 2000). As drought conditions worsened in Idaho in the early 1990s, the level of water transferred through Idaho's established banks declined (as opposed to increasing levels of transfers in California during droughts). A study comparing the experiences of California and Idaho with water banks thus concluded that in Idaho, "from the perspective of the broader society, the banks did not promote the most efficient use of the available resource" (Miller, 2000).

TABLE 6-2 Summary of WAP Water Transactions

Fiscal Year	Total Water Acquired (acre-feet)	Price Range ($/acre-foot)
2001	190,424	60-150
2000	64,995	25-125
1999	232,500	60
1998	91,100	15-700
1997	273,539	15-70
1996	47,152	25-40
1995	101,832	36-50
1994	43,322	50
1993	1,559	34-40
Total	1,046,423	15-700

SOURCE: Available online at *http://www.usbr.gov/mp/cvpia/wap/docs/summary.html*, last accessed June 10, 2004.

The U.S. is not the only nation grappling with the issues of limited water supplies, increasing demands, and environmental concerns. In Australia, for example, the Murray-Darling Basin Commission works in a setting with some parallels to the Columbia River, including an arid climate, an important irrigated agricultural sector, and pressing environmental concerns. In response to environmental stresses on the Murray-Darling River ecosystem, in 1995 the commission's Ministerial Council introduced an interim "cap" on diversions of water from the basin (interbasin transfers from the Murray-Darling are an important source of irrigation water), which was confirmed as a permanent cap in 1997 (see Box 6-1). Lessons from experiences in the Murray-Darling River basin in balancing economic and environmental needs may have relevance for water managers facing similar challenges in the Columbia River basin and across the western U.S.

Water Conservation

Water markets are designed to enable transfers of water supplies among potential users. Typically, the supply of available water is assumed to be fixed over a particular time period. However, it is possible to increase the amount of water available for

> **BOX 6-1**
> **The Murray-Darling River Basin Cap**
>
> The Murray-Darling River basin covers much of southeastern Australia and includes some of Australia's best farmland and some 2,000,000 inhabitants. Diversions of water from this river basin have increased steadily since the 1950s, which resulted in both important economic benefits and substantial changes to the river's flow regimes. Median flows in the lower stretches of the Murray River were reduced to 21 percent of predevelopment flows. The environmental effects of these reduced flows included loss of wetlands, reductions in the number of native fish, and an increase in salinity levels.
>
> In 1995 the Ministerial Council of the Commission produced a report (*An Audit of Water Use in the Murray-Darling Basin*) that confirmed increasing levels of diversions (much of them for cotton production) and attendant declines in ecosystem health. The council determined that the balance between economic and social benefits from water development, and benefits from instream flows, needed to be revisited. The council thus implemented a permanent cap on diversions of water from the basin in 1997. The cap does not attempt to reduce diversions but rather to prevent them from increasing, as it aims to restrain diversions not development. Establishment of the cap marked a substantial change in Australia's water-sharing framework, and it will require considerable adjustments from water users and management entities in the basin. In enacting the cap, the council is promoting a new emphasis on water use efficiencies, reductions in groundwater withdrawals, and a more efficient framework for water trading between states and between individuals. The Australian Bureau of Agricultural and Resource Economics has estimated that more widespread use of water trading in the basin would increase economic output by around $48 million (Australian) annually (for more information on the cap, see *http://www.mdbc.gov.au/natural resources/the_cap/the_cap.htm*, accessed February 16, 2004).

transfer by encouraging water conservation. In many parts of the country, water conservation has emerged as an important source of "new" water supply. In places where available surface and groundwater supplies are fully appropriated or overappropriated, making more efficient use of existing supplies frees up water to serve new demands. Substantial opportunities exist in all sectors to reduce the volume of water used and to decrease adverse impacts on water quality. Partly in order to meet water quality standards, many industries have installed closed systems, which recycle water supplies and consequently reduce the amounts of industrial water used.

Irrigated agriculture has made great strides in increasing water efficiency through means such as the lining of ditches, laser leveling of fields, and more efficient water delivery systems. Incentives for such changes have sometimes derived from legal institutions. The concept of beneficial use includes prevention of waste, and most legal authorities view the concept of beneficial use as useful in encouraging the installation of conservation infrastructure. For instance, as part of the active management areas in Arizona created by the Arizona Groundwater Management Act, beneficial use has been defined as a best management practice. In Arizona groundwater rights are periodically adjusted downward as conservation technologies improve. The program initiated by the State of Washington under the Columbia Basin Initiative appears to be taking an unique approach in that water rights become more secure (i.e., noninterruptible) when better management practices are installed on participating farms.

Conservation infrastructure in agriculture is expensive, and farmers are not likely to make such investments without incentives to do so. Even if conservation leads to better crop yields and reduced pumping costs, the cost of initial investment may be prohibitively expensive. The federal government, through the Natural Resources Conservation Service of the U.S. Department of Agriculture, is providing low-interest loans, cost-sharing arrangements, and other incentives to make such investments more attractive. If farmers are able to transfer or sell conserved water (as is the case in Oregon), conservation investment is a more attractive proposition. As discussed in this report, farmers in California's Imperial Valley have negotiated with the Metropolitan Water District to transfer conserved water to urban users in exchange for financial support in the installation of conservation technologies, and similar strategies would seem to hold promise in the Columbia River basin. It should be remembered that water conservation measures may reduce diversions or losses (e.g., seepage to groundwater through unlined conveyance canals) but that they do not reduce crop physiological needs. The implications of this fact for the quantity and quality of return flows should be considered in discussions of potential water transfers.

Municipal water use is the fastest-growing demand in the West. Urban water pricing, use, and conservation policies can be valuable in helping reduce lower-cost and wasteful uses. For example, a recent study of California urban water use policies concluded that "California's urban water needs can be met into the foreseeable future by reducing water waste through cost-effective water-saving technologies, revised economic policies, appropriate state and local regulations, and public education" (*http://www.pacinst.org/reports/urban_usage/waste_not_want_n ot_exec_sum.pdf,* last accessed March 23, 2004). Building codes requiring low-flow toilets and other water-saving appliances can make a substantial difference in indoor water use. In most cities in the arid western U.S., outdoor watering in the summer constitutes a large water use that can be affected by water conservation policies. Many cities have differential summer and winter water rate structures, with additional costs levied on customers whose use rises sharply in the summer. Also, many cities have increasing block rate structures in which the more urban residents use, the more they pay. Most cities also engage in public information campaigns stressing the scarcity of water and the need for conservation. The urban water conservation literature notes that the artificially low water rates common to most cities undercut conservation incentives. Elected officials are often reluctant to raise water rates. Some experiences suggest that such actions may have political costs; for example, in the 1970s several members of the Tucson, Arizona, city council were removed for sharply raising water rates during the summer. On the other hand, some surveys have indicated that customers have a high willingness to pay for safe, reliable, high-quality water services (AWWARF, 1998; NRC, 2002b). As long as relatively cheap sources of additional water are available for diversion, there is little incentive for urban water utilities to press elected officials to adopt rate increases sufficiently to prompt serious urban water conservation. In establishing urban water use fees, the relations between fees and conservation incentives should be considered.

Adjusting to Water Shortages

The climate of the Columbia River basin is characterized by annual fluctuations in snowpack, precipitation, and streamflow.

Both natural systems (e.g., aquatic habitat) and managed systems (e.g., irrigated agriculture) have evolved in ways to accommodate variations in precipitation and streamflow. In the case of irrigated agriculture, the ability to adapt to changes in water supplies is central to this sector's economic viability. The ability to make such adjustments can provide insights into the consequences of water permitting decisions. That is, what adjustments are available to agriculture if new permits are limited and some classes of water rights (interruptible) are not changed?

One means by which farmers respond to drought is by securing water supplies from alternative sources. For example, in the Klamath basin of Oregon during the 2001 drought, the Oregon Water Resources Department permitted the development of several drought/supplemental use wells (OSU/UC, 2002). California growers have routinely used wells to supplement surface water supplies during drought periods. Continued use of wells in Oregon, specifically during nondrought years, will be limited by growers' ability to obtain permanent water rights for them. Recently drilled wells are generally permitted for use only during declared droughts (ibid.). The use of such wells during nondrought years will also be affected by water quality issues, interference with previously permitted wells, and high operating expenses. In addition, there is evidence of marked groundwater drawdowns during drought periods. Although emergency wells offer important flexibility during a drought, their usefulness during future droughts is thus not always certain.

For farmers (both those who are able to secure additional water through wells or purchases and those who must adjust to reduced supplies), a basic decision in the face of drought is to determine which combination of crops and fields could be successfully planted, irrigated, and harvested under a changed water supply. For example, acres that would not receive sufficient irrigation are usually left fallow. Crop rotation and selection are important management tools used by irrigators, regardless of water availability. Rotation of low- and high-value crops maintains soil productivity, reduces disease, and moderates interannual variability in revenue. In addition, since water requirements differ across crops, if one of the crops used in the rotation requires less water per acre to produce a harvestable yield, some excess water may be available for other crops in the rotation. In fact, rotation patterns and the resulting harvest and income patterns often continue relatively undisturbed, even in drought years.

During a water shortage, a farm's water is typically directed first to high-value perennial crops, such as apples or grapes, or to high-value annual crops such as potatoes. Producers plant high-value crops in fields with reliable water supplies, sometimes despite the presence of inferior- quality soil. During drought periods, low-value crops like wheat and hay are often irrigated with reduced amounts of irrigation water or may not be planted at all (Faux and Perry, 1999). Jensen and Shock (2002) have suggested a set of possible responses to drought by irrigators in the Pacific Northwest:

- Leave some ground idle, applying water first to high-value crops;
- Avoid overirrigating by using evapotranspiration charts to estimate crop water need, soil moisture monitoring equipment, graphing soil moisture readings, and knowing the water-holding capacity of different soils on each farm;
- Know the drought tolerance of different crops and plant according to water availability;
- Implement alternative irrigation methods, such as surge irrigation, on the first irrigation to reduce water loss to deep percolation;
- Switch to sprinkler or drip irrigation for high-value crops like orchard crops, if it is cost effective;
- Change irrigation sets when water reaches the ends of the furrows, rather than at specified times of the day;
- Eliminate deep watering of shallow-rooted crops and employ more frequent irrigations of smaller amounts to keep water in the plants' root zones;
- Use catch basins to capture and reuse runoff.

Historically, irrigators have incorporated flexibility in annual cropping and irrigation decisions to help moderate interannual variability in exogenous factors, like weather and prices. In general, producers manage their crops during a drought year as they do through an average or wet year, with a diverse set of crops, flexible planting, irrigating, and harvesting schedules, and an expectation that low yields during dry years will be offset by high yields during average and wet years. However, prolonged droughts, failure to secure operating capital due to lenders' perceptions of risk, and institutional mandates, such as provisions of the Endangered Species Act, pose special challenges to irrigators. These types of adjustments to reduce water will take on increasing importance as the demand for water from other uses increases.

SUMMARY

The economic value of water in different uses varies widely in the western United States. Equity, intergenerational considerations, and other factors suggest that water should not always simply be allocated to the highest bidder; nonetheless, willingness to pay indices demonstrate that water does provide different types and amounts of economic and social benefits in different uses. The traditional doctrine of prior appropriation in the western United States allocate water rights according to the principles of "first in time, first in right" (establishing a system of seniority of rights), and "use it or lose it" (water rights are open to forfeiture if not beneficially applied). Prior appropriation requires that water be put to beneficial uses, but it does not prioritize water rights based on willingness to pay considerations, or the economic or social return, of water applications.

Water uses and water demands changed greatly across the West during the twentieth century. The "New West" (Riebsame, 1997) features increasing urban populations, changing employment patterns, changing cultural and leisure preferences, an increase in nontraditional economies and employment (e.g., recreation, tourism), and a decreasing economic reliance on traditional activities such as ranching and irrigated agriculture. Traditional sectors remain important, however, in many areas in the West, and there are increasing pressures and competition for often-limited water resources. Some of the pressures for water resources take the form of demands to not divert water from streams but rather allow "instream" flows in place for ecological and related social benefits. The doctrine of prior appropriation did not historically recognize instream flows as "beneficial," but changes in the doctrine in many western states were made during the late twentieth century to recognize the social and economic benefits of instream flows.

The pressures of increasing human population and shifting

social preferences with regard to water resources represent both opportunities and potential conflict in the West. Opportunities exist for water rights holders in traditional sectors, such as irrigated agriculture, to sell water rights for a profit to higher-value uses. Opportunities also exist for traditional users to manage water better—through conservation or better technology—and to sell a portion of this "saved" water. Problems may arise, however, when market-based sales of water between willing buyers and sellers result in third-party effects, which should be carefully guarded against, as these effects can be economically and socially damaging. Conflicts may also arise when traditional users are not interested in selling or when traditional users and newer users vie for the same limited water resources. These conflicts also suggest some roles for governmental bodies in helping ameliorate third-party conflicts and in making decisions about allocations between competing users. Many of these opportunities and conflicts are manifested in the Columbia River basin and along the mainstem Columbia River.

The doctrine of prior appropriation has some flexibility in allowing water rights to be transferred or sold. For example, water rights under prior appropriation are often not attached to land rights and may thus be sold separately from land, which helps effect some water rights transfers and sales from lower- to higher-value uses. Water markets and water banks attempt to increase this flexibility by improving communication and effecting interactions between potential buyers and sellers of water. Although they are not perfect, water markets and banks have demonstrated advantages in producing both flexibility and security in a number of contexts. Market-based programs—several of which have been used to good effect across the West—such as water banks, environmental water accounts, and water rights sales and leases, along with careful monitoring of outcomes, would allow management organizations to learn more about the value (or lack thereof) of these various programs. These market-based measures can also improve incentives for water conservation through better management or new technologies, as conserved water could be sold for profit through markets or banks. These nonstructural water management measures also offer alternatives to traditional means of "increasing" available water (e.g., additional storage reservoirs or diversions). Thus, in addition to helping increase overall social benefits of water uses, these measures hold the prospect of decreasing conflicts over limited water supplies. Water management entities across the Columbia River basin should cooperate on exploring the utility of these measures that can help support the regional economy, but without additional withdrawals of Columbia River water, as the well-being of salmon habitat and salmon is also an integral part of the regional economy. Water conservation measures and means for reallocating water, such as water banks and water markets, should be promoted in a quest to increase "water productivity" and to contribute to a healthy regional economy and Columbia River ecosystem.

As discussed in this chapter, water markets and water banks present their own unique set of implementation and operational challenges. Such programs often require the creation of significant administrative structures, leadership skills, and wisdom in order to ensure that potential buyers and sellers have good information and are aware of each other's demands, and that there are adequate, effective databases that reflect ongoing transactions and that help ensure fair execution of lease and option arrangements. They also require adequate storage and conveyance facilities to store and reallocate water; capital investments in such facilities may also be required. The human resources requirements to ensure the transparency and credibility of such programs may be considerable. Moreover, the wide range of business, economics, and administrative skills necessary for such programs is often not widely available in most natural resources agencies. Successful creation of water markets and water banks thus often holds great potential to identify "new" sources of water and may therefore increase beneficial uses and reduce tensions; but human resources investments to ensure that adequate organizational, environmental, and social frameworks are essential and may be substantial. **The State of Washington and other Columbia River basin entities should continue to explore prospects for water transfers and other market-based programs as alternatives to additional withdrawals.**

7

Water Resources Management, Risks, and Uncertainties

Decisions regarding permit applications for consumptive water withdrawals from the Columbia River involve imprecise calculations and assumptions of salmon's physiological needs, river flows, and present and future amounts of upstream water uses. These decisions must thus consider and balance a variety of imperfectly understood risks. This chapter examines issues associated with managing these risks. It also examines challenges associated with using scientific information in decision-making applications. Part of this study includes the review of several water management scenarios (also listed in Appendix B). Comments on these scenarios are located near the end of this chapter.

RISKS AND WATER MANAGEMENT

A Simple Stream

This example assumes a stream with three users, all under the jurisdiction of one state: User A in the headwaters, User B in the stream's middle reach, and User C in the stream's lower reach. The average instream flow is 15 units of water. For purposes of this example, recognizing that reality is far more complex, it is assumed that salmon need a minimum of 4 units of water. Three possible variations are considered:

Variation 1: User A consumptively uses 5 units of water. After this use, 10 units remain in the stream at the top of the middle reach. User B consumptively uses 3 units of water but wants to expand use to 5 units. The state could permit User B to do so, since 5 units of water would still remain in the river, one more than the salmon "need."

Variation 2: After the new uses contemplated in Variation 1 are implemented, User A wishes to expand use in the upper reach. The state could permit the consumptive use of 1 additional unit in the upper reach without adverse effects to salmon in the middle reach under average or normal conditions. When the upper-basin water supply is less than normal, however, users A and B will both continue their uses until the water available to salmon is exhausted, unless that water is afforded legal protection. Thereafter, User A's junior rights will be curtailed in favor of User B's senior downstream rights. Unless the water requirements of salmon in the middle reach have legal protection, however, the salmon will suffer adverse effects in below-average water years.

Variation 3: Now consider User C's downstream uses that require 5 units (they could be consumptive or nonconsumptive). In a normal water year, User B must pass that much water through the middle reach. Since this "pass-through" water also benefits salmon in the middle reach, User B can still consumptively use 5 units in the Middle Reach (10-5 = 5 units "pass through" to downstream uses). An additional unit of development can occur in the headwaters or middle reach. Beyond that margin, the water needed for salmon will be reduced.

If the stream is wholly within the jurisdiction of one state, these variations can be successfully managed so long as salmon instream flow requirements have legal protection. Such protection can result from a water right or water reservation with its own priority date that is administered along with other priorities on the stream. Legal protection also can result from a regulatory program, perhaps under the Endangered Species Act or under a water quality statute that requires maintenance of a given streamflow. Without legal protection for the water necessary for salmon in the middle reach, however, increases in upstream water uses may eventually encroach upon flow levels required to sustain a salmon population(s).

A Complex Stream

Legal and Water Availability Uncertainties

A more complex situation (and more similar to that in the Columbia River basin) is considered below. Rather than a basin within a single state, the Columbia River is subject to a complex jurisdictional web. User A is no longer a single user whose uses are permitted by a single state. In the Columbia River basin, "User A" is the collective of many upstream governments and entities. Looking upstream, water is currently used in Canada, Montana, Idaho, and Wyoming, as well as Washington and Oregon. There are also potential future uses, such as potential Canadian claims under international law, equitable claims of Montana and Idaho to the waters of an interstate system, and indigenous and reserved water rights claims of upstream Indian tribes.

Water availability in the middle reach is also subject to existing and future downstream claims as well (User C in the previous example). These include claims for sufficient water for navigation, senior water rights for federal reclamation projects, other senior water rights, claims of downstream Indian tribes for instream and diversionary purposes, and equitable claims of Oregon, Nevada, and Utah to the waters they contribute to the interstate system. Other so-called federal regulatory water rights such as sufficient flows for water quality and the protection of listed species under the Endangered Species Act, impose limits on water use both upstream and downstream of the middle reach.

Whether one looks at upstream or downstream rights, present flows in the Columbia River mainstem do not necessarily accurately reflect current legal allocations. In addition to climate trends and variations, actual withdrawals may be augmented by water rights not being currently used and water rights applied for but not yet adjudicated. Although legal entitlement is supposed to be contingent on actual and continual use, water use is not always carefully monitored. Some water rights holders may go many years without diverting their full entitlement. This is important because unless full rights are extinguished for lack of use, they may emerge as significant withdrawals at some unpredictable future time. Further, even in Washington, some surface waters have not been adjudicated. Approximately 160,000 pre-1917 surface water claims and pre-1945 groundwater claims remain unadjudicated statewide, although it is uncertain how many of these claims are contiguous to the Columbia River. Approximately 100 pre-1917 claims for surface water list the Columbia River in eastern Washington as a source.

On the other side of the ledger, illegal diversions may inflate actual withdrawals from the river. The National Marine Fisheries Service 2000 Biological Opinion suggests that controlling illegal use of water at U.S. Bureau of Reclamation irrigation projects could substantially reduce streamflow depletions. This opinion was pursuant to a 1994 Inspector General's report that detailed unauthorized uses within bureau projects, some of which are situated on the Columbia River. Although the Bureau of Reclamation is aware that unauthorized uses occur, it does not now have a dedicated program or schedule in place to address and resolve all instances of unauthorized use. The bureau has undertaken some efforts, in this realm, including geographic information system (GIS) mapping of the Columbia River Project. Onsite review may be necessary to accurately determine the extent of unauthorized uses, which would require staff and related resources. A report issued by the bureau warns that farmer resistance could make it difficult to quantify unauthorized uses and that efforts to limit unauthorized diversions must be cautious and collaborative (U.S. Bureau of Reclamation, 2000). The bottom line is that current flows of the Columbia River do not present an accurate picture of legal entitlements to withdrawals.

The Columbia River system can no longer be managed under a simple set of priorities. A legal inventory may tell only part of the story. Canada has treaty and equitable rights to the river. Other Columbia River basin states also have equitable apportionment claims. The water rights of these sovereigns, although impossible to predict, constitute "first" claims on the river. In addition to their fisheries rights, many of the basin tribes have reservations with arable land. Under federal law, many of these tribes have reserved water rights for irrigation and other "permanent homeland" purposes. The priorities of these rights may vary depending on evidence of aboriginal use, treaty entitlements, when lands were acquired, and many other factors. State-law water rights for reclamation projects and other uses in Washington and other states may not be fully developed; and, short of a general basinwide stream adjudication, it is difficult to determine how much additional water use is authorized and, if developed, whether priorities of these uses will relate back to the original dates of filing or appropriation. These legal uncertainties exist against a backdrop of variable water supplies. During normal climatic situations, precipitation may vary considerably from year to year. And, as explained in Chapter 3, possible climate warming across the basin may portend increasingly erratic patterns of streamflow and water availability.

Risk and Uncertainty Involving Endangered Species

The existence of threatened and endangered species, such as salmon and steelhead, can further complicate water management decisions. A previous National Research Council committee addressed risk assessment in the context of the Endangered Species Act (NRC, 1995). That report identified the two types of risk addressed by this statute: the risk of species extinction and the risks of potentially unnecessary expenditures of money and curtailment of resource use given the uncertainties about the risk of extinction. The report enumerated major factors that appear to influence the risk of extinction and discussed the difficulties of estimating the risks of extinction, many of which are presented in considering potential new diversions from the Columbia River. In addition, other NRC committees have examined issues of the use of risk analysis in water management and risk communication (NRC, 1996, 2000). Based on definitions and applications in those reports, this report defines *risk* as the probability that some undesirable event occurs, as well as the combination of that probability and the corresponding consequence of the event. *Uncertainty* is used to describe the lack of sureness about something.

A key risk in many situations in which species face extinction is the relatively small population size of the species. In small populations, even random demographic or environmental changes can have large consequences for species survival. Catastrophes such as drought or fires can suddenly reduce population numbers. In small populations, genetics can also be a factor as "mildly deleterious genes, previously kept at low frequency by natural selection, can rise to high frequency by change" (NRC, 1995, p. 133). The species' ability to adapt genetically to environmental changes is also diminished because genetic variation, the key to species' adaptation, can be compromised by reductions in population size (ibid., pp. 134-135). The report noted, for example, that "Populations need about 1,000 individuals to maintain their genetic variation" (ibid., p. 135; the report also noted that this actual number depended on the biology of the organisms involved). Applying these findings, random demographic and genetic changes are likely not primary risk factors for the threatened and endangered salmon runs in the Columbia River. These factors are more detrimental in populations of fewer than 1,000 individuals, and the Columbia River salmon runs, although in jeopardy, are more numerous. Habitat fragmentation is another important risk factor and, because of the many physical alterations of the Columbia River system, of greater concern to salmon survival. The report also noted that the effects of even minor detrimental changes in specific habitat areas may accumulate over time—an especially relevant observation in the context of this report's considerations of how Columbia River water withdrawals affect salmon survival rates. As the report states:

> Not enough is known about cumulative effects and threshold points. . . . When considering the probable effects of incremental human activities, it is reasonable to assume that additional activity means additional risk, but we rarely know whether the relationship . . . is linear or whether there might be critical levels of activity above which the risk of extinction increases dramatically. (p. 156).

Compounding of Uncertainties

All these legal, economic, biological, and water availability uncertainties intersect in water permitting decisions concerning the Columbia River middle reach. There are many legal and economic uncertainties regarding how much additional water will be consumptively used upstream by Canada, other states, and tribes. While downstream uses help "pull" water downstream for salmon, similar legal and economic uncertainties exist about the growth of these downstream uses. Ocean conditions also influence levels of salmon returns to the Columbia River. Climatic uncertainties confound precise predictions of how much water will be available to use throughout the basin.

The "risk-based" nature of permitting in the Columbia River middle reach is suggested by Table 7-1. The rows of this table represent different assumptions about water availability in a given year, with high flow conditions (risk value 1) presenting lower risks to salmon and low flow conditions (risk value 3) presenting higher risks. The columns display different risk assumptions about the extent of upstream consumptive use in that year. If little additional water has been allocated for consumptive use in the upper basin, the risk to salmon in the middle reach is low (risk value 1). If much upstream allocation has occurred, the risk to salmon in the middle reach is high (risk value 3). The shaded cells in Table 7-1 indicate the products of the interaction of water availability risks and consumption risks. The darker cells suggest high risk to salmon in the middle reach of the Columbia River in circumstances of low water availability and high levels of upper basin water development (i.e., a high risk of low flows compounds the situation of having high levels of use in the upper basin).

Table 7-1 depicts risks to salmon presented by the varying relationships between river flows and upstream consumption. Similar tables could be drawn for all the other variables (e.g., temperature, habitat) that affect salmon survival. The problem for water managers is that the risk factors in such tables combine with one another. That is, the risk to salmon viability is a composite of all these individual risk factors. If managers are confident in scoring all these risk factors low (lighter-color cells), additional permits can be issued with assurance that impacts on salmon will be minimal. However, if managers score many or all the individual risk factors high (darker-color cells), additional permitting could affect salmon adversely. Perhaps an even greater challenge is that seldom are these varying risks to salmon quantified as precisely as suggested in the table. In nearly all cases, risks are only partially understood and entail some qualitative understanding and a need for professional judgment in decision making.

TABLE 7-1 Hypothetical Risks to Columbia River Salmon*

		Levels of Upstream Consumption		
		Low (1)	Medium (2)	High (3)
Given Flows for Specific Water Year	Low flows (3)	Risk = L x L = 3	Risk = L x M = 6	Risk = L x H = 9
	Medium flows (2)	Risk = M x L = 2	Risk = M x M = 4	Risk = M x M = 6
	High flows (1)	Risk = H x L = 1	Risk = H x M = 2	Risk = H X M = 3

* Risk values range from 1 (lowest) to 9 (highest).

COLUMBIA RIVER MANAGEMENT DECISIONS

Anticipated Permitting Decisions

Under optimal conditions, a permitting agency could make confident predictions of existing and anticipated water use, especially above the reach in which additional permitting is planned. The permitting agency would also have reliable estimates of future water availability and the distribution of those flows throughout the year. Potential permitting decisions for the middle reach of the Columbia River, however, present a less than optimal situation since, from a legal perspective, existing and future upstream water uses are difficult to determine and water availability is subject to variability at various timescales. Further decreases in flows or increases in water temperature will increase the risks associated with managing water resources and salmon and are likely to reduce survival rates. The confluence of some, or all, of the many factors that threaten to reduce Columbia River flows poses serious risks for salmon, many of which are endangered. Given the current setting and likely future climatic and other trends, additional water withdrawals from the river during seasons characterized by low flows (particularly in drought years) will pose additional risks to salmon survival, which should be considered in decisions regarding potential future Columbia River withdrawals during low flows.

Interjurisdictional Cooperation

Many of the risks that confound permitting decisions in the middle reach of the Columbia River result from upstream uncertainties that Washington State has little control over. An inventory or model of water rights on the Columbia River cannot be reliably created because the extent of many of the largest rights cannot be determined until adjudications, other litigation, or settlements are completed. An interjurisdictional water organization is one means that could help better manage or even reduce uncertainties. Such an organization could provide a forum for improving information and assessing the consequences of major management actions on the Columbia, as well as providing a broader setting for discussion and learning. Such an organization should include the basin sovereigns—the Canadian and U.S. governments, U.S. basin states and Canadian provinces, and Indian tribes. The body should establish a means to incorporate and discuss scientific input. This body should establish a threshold(s) volume of proposed new withdrawals that would be likely to concern more than one government. For instance, any proposed new use of water of more than an agreed-on amount could be considered presumptively suspect and would have to be referred to the interjurisdictional organization for deliberation. The organization's decision rules might require hearings and a complete record on the basin consequences before the project could continue. The rules might also require the organization's approval before the permit could be issued.

Incremental Actions and Adaptive Management

Consideration of water permit applications in the State of Washington takes place in a contentious and turbulent science and policy context. The body of scientific knowledge of Columbia River salmon is complex and incomplete, and there are competing scientific theories regarding some of the relationships between salmon and environmental variables. There are also many decision makers with differing goals, a situation noted in a 1989 article on Columbia River management, "A . . . problem is the large number of hands on the steering wheel" (Lee, 1989).

The setting of multiple political jurisdictions, competing

stakeholder groups, endangered species, a complex ecosystem, and a large but imperfect body of scientific knowledge is not unique to the Columbia River and in fact is characteristic of many major U.S. river systems. In an effort to implement management regimes that help reduce stakeholder disputes and that strike a more amenable balance between legal obligations and authorizations, many management agencies in the United States (and abroad) are exploring the prospects of "adaptive management" strategies. Adaptive management has its foundations in many different fields, but its theories and concepts were formalized by ecological scientists in the 1970s (Holling, 1978; see also Gunderson et al., 1995: Gunderson, 1999; Lee, 1993, 1999; Walters, 1986). Elements of adaptive strategies include:

- An explicit recognition of uncertainty and the need to learn more about coupled ecological-social systems in order to enhance learning and reduce uncertainties;
- Recognition that adaptive management entails a process, not a final answer or a series of management "endpoints" to be rigidly pursued;
- Learning while doing. Adaptive management does not postpone management actions until "enough" information is available (Lee, 1999). It seeks management actions that can be reversed in light of new information and actions that can help improve ecological understanding while also meeting economic and environmental needs. Adaptive management is not "trial-and-error" management, but rather entails carefully designed management actions, with purposeful monitoring of outcomes in a structured learning process;
- Flexible, incremental actions that enhance learning and that seek to avoid catastrophic error;
- A means of gathering information on environmental and economic outcomes of management decisions;
- A vision or a model of the ecosystem that is being managed (Walters, 1986). This vision or model provides a baseline for defining surprises. Surprises and other new information help increase knowledge and understanding of the system (Lee, 1999);
- Organizations that can learn from new information and policies that can be adjusted in light of new information;

- A collaborative structure for stakeholder participation. Participants should be willing to negotiate, try a variety of temporary measures, and evaluate promising measures before they are implemented. Adaptive management does not seek to eliminate differences of opinion or conflict but rather to provide a framework for their discussion. Adaptive management is not a substitute for willingness to compromise and give-and-take, however, and unless stakeholders are willing to agree on basic questions or lines of inquiry to be pursued by an adaptive approach, formal adaptive management will be inappropriate. Well-managed conflict can be a resource for new ideas and approaches and mutual learning, but one cannot manage adaptively in the absence of stakeholder flexibility (Gunderson, 1999).

An adaptive management approach would encourage Columbia River basin entities to move forward incrementally and try a variety of approaches for better understanding and managing risks and uncertainties. Decisions and policies should promote flexibility while their outcomes are being evaluated and better understood. A broad range of stakeholder groups from across the Columbia River basin should be engaged in crafting these decisions.

A variety of approaches to meet water demands in the middle reach might be explored. Chapter 6 identified several economics based alternatives, such as cost-shared water conservation improvements, reallocating existing uses, water banks, and water transfers. Most or all of these types of measures could be implemented incrementally and could be amenable to change as new economic and environmental information is gained. Adaptive management aims to yield better information about ecological, economic, and policy conditions, reduce uncertainties, and engage participants in a collaborative learning process aimed at solving complex problems, such as Columbia River management. The following section discusses the use of scientific information in decision-making contexts that are laden with uncertainty.

Science and Decision Making

A vast amount of scientific research on Columbia River salmon has been conducted over a period of several decades.

The resulting body of knowledge provides a broad understanding of salmon life cycles and histories, physiological characteristics of salmon, and environmental variables important to salmon survival. As explained in Chapter 4 and in other sections of this report, Columbia River salmon inhabit and travel through extensive oceanic and riverine systems during their life cycles. The size and the complexity of these systems, and the biological complexities of salmon, frustrate attempts to understand any of these factors with high precision and certainty. Substantial resources have been devoted to investigating Columbia River salmon, and today these fish species are one of the most intensively studied in the United States, if not the world. Although scientific understanding of the salmon has improved over the decades, perfect understanding of all factors and relationships that affect salmon life cycles is beyond current and foreseeable future scientific means.

More precise scientific information regarding salmon behavior, environmental influences, and rates of survival could, over time, no doubt be obtained. However, significant resources are now being devoted to this pursuit, as federal and state scientists and scientists from universities and regional consultancies are involved in extensive salmon research programs. One task pursued in this study concerned the identification of knowledge "gaps" and "scientific information" needed to develop comprehensive strategies for recovering and sustaining listed species and managing water resources to meet human needs (see Chapter 1). This task, however, presupposes that sound management strategies can be devised only when scientific "gaps" are filled and that it is possible to determine a priori the scientific information that will lead to better management decisions. Such suppositions do not reflect contemporary natural resources management realities and the relationships between scientific information and decision making processes.

Identifying the additional scientific information that will prove useful for management is not strictly an issue of scientific inquiry but also a matter of policy-making processes. Scientists are often expected to provide specific answers for use in decisionmaking and policy making. This may place an undue burden on scientists, however, especially given the uncertainties and risks that revolve around such issues as Columbia River salmon. Science is a key component in these decisions. But rather than looking to science to provide information in strictly a one-way direction, decision makers should collaborate with scientists in a two-way process in which management actions are taken in the face of some inevitable uncertainties, with an eye to learning more about the system(s) at hand. Progress toward "comprehensive" management cannot be accomplished through scientific inquiry alone, but rather requires stakeholders and management agencies to work with scientists in a collaborative learning process, such as that framed by adaptive management principles. As stated, Columbia River salmon management is an exceedingly complex public policy and science issue. The creation of "comprehensive" strategies that reduce tensions, protect and enhance salmon, and respond to shifting human needs will likely require an approach that mirrors these complexities, as suggested in the following passage by Lee (1989):

> Sustainable development of the Columbia River basin requires managing an ecosystem the size of France. If there is to be a sustainable Columbia, it will be a place governed by rules that approach the complexity of ecological interaction.

In cases where there are sharp conflicts and differences of opinion, management agencies may understandably be reluctant to take decisive actions in the face of uncertainty. Such a stance, however, may contribute to the buildup of tensions among stakeholders. In these settings an adaptive approach may be useful. Adaptive management does not wait until "enough" information is available but recognizes that gaps are inevitable, that data collection is expensive and time consuming, and that there are sometimes problems requiring decisive actions in the face of limited information. The approach seeks to create flexible management regimes through a collaborative science and management process. Maintaining flexibility of management decisions to the maximum extent possible is essential. Additional scientific research on Columbia River salmon should continue. Better information on flow-survival relationships, for example, can reduce uncertainties and contribute to better management decisions. Scientific inquiry on the salmon should be allied with policy making and stakeholder participation in an iterative, interactive process. Adaptive management can help participants better understand the ecosystem. However, it requires willingness among participants to find common ground and a political will to act in the face of uncertainties.

Adaptive management is not a foreign concept in the Pacific Northwest. The Northwest Power and Conservation Council has sought to manage Columbia River fish and wildlife resources under an adaptive management paradigm. The first serious attempt at implementing adaptive management principles began in 1986, and the process has proceeded with a variety of initiatives (Lee and Lawrence, 1986; Volkman and McConnaha, 1993). Although adaptive management holds promise for improving understanding of flow-survival relationships in the Columbia River, the political setting is highly contentious, economic interests and values are substantial, and management responsibilities are dispersed among many entities. Its implementation may also be inhibited by the Endangered Species Act, as the adaptive management paradigm of accepting risks and occasional mistakes as part of a learning process runs counter to the ESA's aversion to risks. Management actions aimed at helping improve understanding of flow-survival relationships may indeed, as Volkman and McConnaha (1993) have asserted, "kick off a new round of battles." There can be no denying the political challenges and scientific complexities that attend adaptive management principles. But the complexities of managing Columbia River flows and salmon defy simple solutions and will likely require a management paradigm of similar complexity. Although stakeholders may currently share little common ground, it is important to explore innovative ways to improve on the current management regime. Although it does not represent a panacea, adaptive management offers a systematic, collaborative learning and management process as an alternative to allowing decisions to be made through court litigation and decrees.

THE MANAGEMENT SCENARIOS

The Washington State Department of Ecology provided five management scenarios for evaluation within this study under item 5 of the Statement of Task ("Evaluate the effects of proposed management criteria, specific diversion quantities, and specific features of potential water management alternatives";

see Chapter 1). The scenarios as provided are given in Appendix B, but because aspects of the scenarios included many details (and are not entirely transparent), they are paraphrased below.

Scenario 1

In this scenario (as in all the scenarios), it is assumed that water can be used between the Canadian border and Bonneville Dam. New permits would be issued to water users in Washington over a 20-year window (the start date is not specified) up to a total of 1 million acre-feet. Of that total, 220,000 would be allocated to the Columbia Basin Project. In addition to the million acre-feet made available to Washington State users, 427,000 acre-feet of instream flow from the Snake River would be "legally recognized throughout the Washington State reaches of the Snake and Columbia Rivers" and "600,000 acre-feet would be recognized as necessary to meet the water resource needs of the state of Oregon."

This scenario implies that about 1,600,000 acre-feet would be used for out-of-stream uses (1 million in Washington and 600,000 in Oregon) and 427,000 would be devoted to instream flow. In addition, permits that currently are interruptible when streamflow reaches a predetermined level could be converted to uninterruptible status if the owner demonstrates "that current water use conforms to state-of-the-art water use efficiency practices." "Uninterruptible water rights" are pre-1980 state law water rights that have priority over main stem instream flow rights established in 1980. Other pre-1980 water rights based on federal law also have priority over these instream flow rights. "Interruptible water rights" are post-1980 state law water rights that, under certain low flow conditions, may be curtailed to protect mainstem instream flow rights. Additional uninterruptible water rights that would not be curtailed under low flow conditions to protect mainstem instream flow rights are proposed.

All new water rights issued would also require state-of-the-art efficiency and would be metered. The Department of Ecology would periodically assess the management program and use scientific information to accommodate changes in knowledge, with formal reevaluations at years 10 and 20. Finally, the department would seek partners to establish a "functioning water market or 'water bank' for the mainstem of the Columbia River to facilitate a more efficient allocation of existing water resources in the basin."

Scenario 2

Scenario 2 is similar to Scenario 1, with the following differences:

1. All new permits and previously interruptible rights converted to uninterruptible status would be charged $10 per acre-foot per year to support additional efforts toward "salmon health and recovery." The proceeds would be used to acquire water for instream flow in low-water years and to make habitat improvements in the mainstem and tributaries. The money might also be used to explore the development of storage projects (these storage projects are not described in detail. Because new storage facilities on the Columbia River mainstem are not a viable option, the implication is that additional storage would be gained by new dams on tributaries; by the creation of new reservoirs to be filled by water from the Columbia River; or other methods, all of which would require additional water withdrawals from the Columbia River mainstem).

2. Of the new permits totaling up to a million acre-feet allocated to users in Washington, 300,000 acre-feet would not be issued until existing users had demonstrated that "conservation investments were in place for a majority of users on the mainstem."

Scenario 3

Scenario 3 is identical to Scenario 2 except that the charge for new permits and for changing interruptible permits to uninterruptible status would be $20 per acre-foot per year. In addition, the Department of Ecology would provide financial support for new conservation measures.

Scenario 4

This scenario would not allow any new water to be removed from the Columbia River for out-of-stream use by Washington users. New water rights would require "direct mitigation in the mainstem of the Columbia River." All new water rights would require offset water to be obtained through water-right changes and transfers, conservation, or use of new storage. Existing interruptible water rights could be converted to uninterruptible status by payment of $30 per acre-foot per year. The money so obtained would be used to acquire water rights to benefit salmon populations.

Scenario 5

This scenario assumes "that the current existing rule governing water resources of the Columbia River" would continue. The current rule includes a moratorium on all new permits, however, and this scenario allows for new permits. Each new permit would be issued only after consultation with fish and fishery managers (e.g., Washington Department of Fish and Wildlife, tribes, NOAA Fisheries) and whether and to what degree mitigation would be required would be decided for each permit individually as a result of the consultation with fish and fishery managers. The upper limit, if any, of the total new water permits that could be issued is not specified.

Evaluation and Commentary

In general, the adoption of concepts related to adaptive management, such as periodic review and adjustment of the program and monitoring, and market-based conservation strategies such as conservation, the use of water markets (or "banks"), and charging for water rights, is commendable. As presented in these scenarios, however, those programs are discussed at only a general level, which precluded deeper investigation and more detailed comments.

A pervasive aspect of the scenarios is the lack of a comprehensive basinwide assessment of water uses and needs as a context for evaluating permit applications. Small (relative to the flows of the Columbia River) withdrawal and permitted volumes will have only small, if not minuscule, effects on the water budget of the basin as a whole. All water uses accumulate, however, both in Washington and elsewhere along the mainstem, as well as the along tributary streams. If future demands for water increase (which seems highly likely given recent and projected demographic and economic trends), the accumulation of risks to salmon survival will be all the greater (given the variety of risks that affect salmon survival, assigning precise and credible levels of risk to changes in flows and temperature is extremely difficult). These effects would be magnified by reductions in low flow that could attend prospective climate warming as well as during periodic unfavorable ocean conditions. The lack of a comprehensive basinwide management structure hampers the ability to make comprehensive judgments (both in time and over space), and it supports this report's recommendation for creating a basinwide framework for coordinating water use data and strategies.

- *Conversion of interruptible to uninterruptible water rights (Scenarios 1-4).* Conversion of interruptible water rights to uninterruptible status makes adaptive responses more difficult. Interruptible water rights are interruptible so that at times of scarcity, instream flows can be protected. Making any out-of-stream right uninterruptible reduces flexibility to retain water in the river when salmon need it most—during periods of high demand and low flows.

The conversion of water rights to uninterruptible status will decrease flexibility of the system during critical periods of low flows and comparatively high water temperatures. Conversions to uninterruptible rights during these critical periods are not recommended.

- *Revaluation at 10 and 20 years (Scenarios 1-3).* The idea of reevaluating the scenarios periodically is excellent. For this reevaluation to be meaningful, however, the program needs to be designed so that any aspect of it could be undone (reversed) if the evaluation calls for such a reversal. No evidence is provided of any such reversibility. Instead, the result will be *decreasing* reversibility by allowing for some interruptible water rights to become uninterruptible. In some cases, more frequent reevaluations might be necessary. In addition, criteria for assessing the state of the art of efficiency measures are not described, and the responsibility for making that evaluation is not specified. There also is no requirement for periodic reevaluation to take advantage of improvements in water use technologies and innovations.

- *Monitoring and metering (Scenarios 1-3).* Monitoring for compliance with standards and metering are excellent ideas and could be accomplished consistent with this report's recommendation for comprehensive basinwide water management. Such efforts will require resources, however, and an estimate(s) of the budget and personnel required to perform such monitoring would thus be useful.

- *Charge for water rights (Scenarios 2-4).* The disadvantages of uninterruptible water permits were considered in this study, and it was concluded that allowing new uninterruptible permits to come into existence, either through conversion or de novo, would decrease the ability of water organizations to manage risks attached to decisions such as the granting of water use permits.

Charges for water rights in this scenario appear to be arbitrarily selected and out of proportion to the probable costs of mitigation and the value of water to the users. For example, the scenarios specify charges of $10 to $30 per acre-foot per year to be used (among other things) to acquire mitigation water in low-water years. This scenario thus proposes increasing the priority of a water permit for $10 to $30 per acre-foot per year and using the money to buy water for what could be several times that amount.

- *Water markets (Scenarios 1-4 and perhaps 5).* As discussed in Chapter 6, water markets, water banks, and other such market-based mechanisms offer potential improvements over existing systems of water allocation. However, restricting markets only to the Columbia River's mainstem, and only to Washington, is narrowly construed. The Department of Ecology already allows for 600,000 acre-feet per year to be used by Oregon in its assumptions, but no allowances are made for uses by Idaho,

Montana, or British Columbia or by tribes. Efforts toward developing water markets should be complemented with efforts to evaluate third-party effects and to design proposals for compensating users indirectly harmed in water rights transfers.

- *Structural storage measures (Scenarios 2-4 and perhaps 5).* A lack of specificity in this scenario inhibits the ability to comment extensively on it. It implies that tributaries are to be used for additional storage (which may have negative consequences for salmon), but the habitat and condition of tributaries are of critical importance for Columbia River water quality and for survival of salmon that use the river. Tributaries should thus be considered for protection and mitigation as well.

- *Scenario 5.* This scenario is not clearly specified. It is not a "no action" scenario, which would entail leaving in place the current moratorium on new permits. Although the idea of consultation with fishery managers is good, no mention is made of criteria for the evaluation, how the results of the evaluation might be enforced, who decides how much mitigation is needed, and what, if any, limits might be placed on new permits.

"Mitigation" measures are suggested in most of the management scenarios. Although the idea of "mitigating" impacts is attractive, the reality of most mitigation measures is that they are not well coordinated; that is, a management agency may attempt to offset harmful impacts of water withdrawals in one part of a river system with mitigation measures (e.g., ecosystem restoration) elsewhere. The ultimate outcomes of such varying actions, however, are difficult to accurately predict, measure, and compare (if indeed they are ever measured and meaningfully compared, which they often are not), thus making it difficult to determine if "mitigation" was actually achieved.

SUMMARY

Columbia River basin water management decisions entail varying degrees of risk to salmon survival. These risks are a function of both the magnitude and the timing of management actions, such as water withdrawals. For example, additional water withdrawals during low-demand periods pose smaller degrees of risk than similar withdrawals during periods of high demand. Decisions are confounded because levels of risk are often understood only on a broad qualitative level. Not only are key variables typically unquantified to a high degree of accuracy, the nature of interactions between key variables is often poorly understood. Some decisions may thus have only limited effects and be made well within a given range of tolerance, while some may result in critical thresholds being exceeded, without a clear understanding of these different impacts.

In this context of uncertainty and varying degrees of risk, it is important that management and policy decisions promote flexibility, and even an appropriate degree of reversibility, in the event of future unforeseen and dramatic consequences. Examples of means by which risks might be managed include organizational learning strategies (which could employ ex post evaluations to learn from successes and failures), interjurisdictional cooperation (which would encourage entities to communicate to ensure that potential gains possible through innovative strategies are not foregone because such strategies are not being employed across an entire watershed), and incremental actions (examples of which include smaller-scale, short-term, and reversible policies). Adaptive management is a strategy that integrates many of these examples. It is not a new concept in managing Columbia River basin fishery resources, and experience and successes with the concept to date—in the Columbia and elsewhere—are limited. The concept does not represent a neat and easy solution to managing the basin's fisheries and water resources, and some may be quick to dismiss it because of its complexity or difficulties in implementation. It should be kept in mind, though, that the exceptional complexity of Columbia River salmon management is likely to entail a similarly complex management framework is to be sustainable and equitable. More scientific information on salmon will not necessarily lead to the resolution of disputes or to better management decisions. **Sound, comprehensive management strategies for Columbia River salmon will depend not only on science but also on a willingness of elected and duly appointed leaders and managers to take actions in the face of uncertainties.**

Sound management strategies will also require a process in which managers and elected officials help frame scientific investigations and inquiry. The scientific knowledge of Columbia River salmon, while as extensive as for any other fish species in the world, is still imperfect. Improvements in salmon habitat and return rates will require a willingness to use existing scientific knowledge to address some of the factors that scientific research suggests have led to their declines. A process in which scientists monitor outcomes of management actions and provide feedback to stakeholders and decision makers, who then adjust management actions accordingly will be instrumental in helping understand how additional scientific research can best support management decisions. This process is generally referred to as adaptive management.

The management scenarios prepared in connection with this study contain some elements that would promote organizational flexibility and have some commonalities with adaptive management strategies that are being used across the United States and in other parts of the world. Although programs such as water banks, water markets, incentives for water conservation, and better metering of water use were presented only very generally and therefore could not be evaluated in greater detail, they tend to support greater water management flexibility and merit careful consideration. Such efforts could meet with resistance from users who have little to no incentive to implement them. The situation calls for creative programs that provide incentives for water users to decrease uses or that identify alternative sources of water supplies.

The State of Washington must consider several variables in making decisions and trade-offs regarding water withdrawal permit applications and the protection of salmon populations. Those variables, which include flows, temperature, and salmon's biological and migratory features, are only imperfectly understood and interact in complex ways. Scientific information can reduce uncertainties, but rarely can such uncertainties be eliminated, especially with regard to issues as complex as Columbia River salmon management. In such settings, decision makers must exercise some degree of professional judgment in balancing a variety of risks and uncertainties. Given the uncertainty of outcomes of these types of decisions, it is important to promote flexible decision-making regimens that can be adjusted as new knowledge is gained.

As this report has discussed, Columbia River salmon are today at a critical point. The basin's salmon populations have long been in a steady decline, and scientific evidence demonstrates that environmental thresholds important to salmon, such as water temperature, are being reached or in some cases exceeded. Salmon are especially imperiled during critical periods of low flows, high demand, and higher temperatures. The risks involved in this context include additional reductions of salmon populations, extinctions, and violations of the Endangered Species Act, as well as risks to other users of the system—such as irrigation farmers—whose water demands may conflict with instream flows needed for salmon and aquatic habitat.

The ultimate decision as to whether to issue additional water withdrawal permits from the Columbia River and nearby areas is one to be resolved by duly elected officials and their appointees in the public policy arena. But in this setting of high risk and uncertainty, if additional permits are issued, they should be issued within a framework that seeks to increase the flexibility of water management systems and organizations. Efforts to enhance flexibility are especially critical given that so many social and physical trends in the Columbia River basin—such as potential additional water claims from tribal lands and other upstream areas, human population growth, and possible climate warming—point to possible reduced water supplies during critical periods and increased risks associated with salmon management. **Decisions regarding the issue of additional water withdrawal permits are matters of public policy, but if additional permits are issued, they should include specific conditions that allow withdrawals to be discontinued during critical periods. Allowing for additional withdrawals during the critical periods of high demand, low flows, and comparatively high water temperatures identified in this report would increase the risks to survivability to listed salmon stocks and would reduce management flexibility during these periods.**

Water permitting decisions made by the State of Washington, as well as by other basin entities, are made with little consideration or obligations of their upstream or downstream implications. This fragmented decision-making basis is a barrier to better water management and a barrier to a more comprehensive and coordinated approach for managing the risks and uncertainties that attend Columbia River salmon management. The Northwest Power and Conservation Council and its predecessor organization, the Northwest Power Planning Council, have served as key entities for promoting cooperative basinwide Columbia River management for over 20 years. The council has accomplished many good things, and adding a responsibility to consider water permitting decisions to its mandate may seem consistent with its natural resources management duties. But trying to integrate these functions in an existing entity could entail complications and drawbacks. A basinwide forum for considering water withdrawal permit applications above a given threshold would provide regional consideration of the systemwide implications of a proposed diversion. This forum need not entail anything binding other than an obligation to refer the applications. At a minimum, proposed diversions would be subjected to professional and public scrutiny, magnitude of risk, possibilities of mitigation, and systemwide equities. A basinwide forum for considering withdrawal permit applications would enhance unified water management across the Columbia River basin. **The State of Washington and other basin jurisdictions should create a joint forum for documenting and discussing environmental and other consequences of proposed diversions that exceed a specified threshold.**

8

Epilogue

The Columbia River basin is a vast hydrological system subject to large and often unpredictable physical, biological, and human-induced changes. Despite the construction and operations of the Federal Columbia River Hydropower System, the river's flows still vary on many different timescales and often in ways that are not fully predictable. In addition, prospective future changes in climate are likely to affect seasonality of flows as well as water temperature. Additional diversions from existing projects and users, as well as additional demands from human population growth (currently increasing and highly likely to continue), are likely to diminish streamflows.

Columbia River salmon populations have been affected by a variety of human activities and have declined over the past century. The declines have been steady but have also exhibited considerable variability, with occasional years of low returns and occasional years of abundant returns, such as those witnessed in the early 2000s. The long-term decline of salmon populations, especially wild fishes, however, is undeniable. Documented increases in Columbia River water temperature are approaching, or have exceeded, thresholds of physiological importance to many salmonid stocks. Migratory behavior and survival rates of salmon are also affected by low river flows. This situation is especially troubling because of prospective future climate warming (which could entail not only higher water temperatures but also further decreases in low flows) and demands for additional diversions of Columbia River water during low-flow periods. Further increases in water temperature and further reductions in low flows would exacerbate risks to salmon survival. As this report has noted, the effects of prospective additional withdrawals in July (234,000 acre-feet) could be substantial. July is a period of high demand for Columbia River water. The upper end of the range of prospective additional withdrawals considered in this study would increase July withdrawals from their current value of roughly 6.8 percent of mean Columbia River flows to roughly 8.6 percent. Under *minimum* July flow conditions, the effects would be greater: the upper end of the proposed range of diversions would increase current July withdrawals from roughly 16.6 to 21 percent of Columbia River *minimum* flows.

The seasonality of Columbia River flows and changing demand patterns for additional water from various users in different parts of the river basin suggest that sound water management decisions require a comprehensive basinwide water management scheme. Ideally, the management framework would have the flexibility to respond to the seasonality of Columbia River flows and have the flexibility to responsibly transfer water from lower-value to higher-value uses. Increased flexibility in managing the Columbia River will require greater emphasis on nontraditional approaches to augmenting water supplies, such as water marketing and water transfers, and greater cooperation of political entities across the basin. These market-based programs may require capital investments in physical infrastructure and human resources investments in experts with skills in fields such as finances, marketing, and public administration. Programs such as water transfers, groundwater banking, and other measures to increase the efficiency of water use hold promise in helping sustain the regional economy in ways that do not require ever-increasing water withdrawals. Although water uses across the basin should not be simply channeled to the highest bidders for water, such measures hold promise for helping support both economic and environmental goals and should be carefully considered.

A key problem in managing the basin's water is that water permitting decisions are currently made in a piecemeal fashion, with little to no consideration of their effects on other users or their degree of consistency with other decisions across the basin. If water resources and risks to salmon survival are to be better managed, Columbia River water permitting decisions must be made in a more holistic fashion, with consideration of how additional diversions would affect other users and sectors across the entire river basin. A joint forum composed of Columbia River basin entities would allow for more accurate inventorying, monitoring, and enforcement of existing water rights. There is also a need for stronger efforts toward water conservation and market-based management strategies, which could help reduce present tensions related to competition over water supplies. Many of these types of nontraditional means for augmenting water supplies have been applied to good effect in some water-short areas of the West. Their prospective applications in the Columbia River basin should be carefully explored.

Water withdrawal applications and permitting decisions are highly contentious in both the State of Washington and other parts of the Columbia River basin. Inflexibilities in traditional western U.S. prior appropriation doctrine have contributed to these tensions. A greater degree of flexibility in traditional water permitting and rights processes is paramount to better water management and to decreasing tensions and conflicts in the basin. This report recommends implementation of a joint basinwide water management forum and the pursuit of nontraditional water marketing and conservation strategies. A water permitting and rights process that more explicitly recognizes seasonality of flows should also be devised. Decisions regarding the granting of new water rights are issues of public policy, but additional water withdrawals during the critical high demand and low-flow periods discussed in this report will increase the risks of survival to listed salmon stocks. It will also decrease the flexibility of management institutions to allocate water between different uses in critical low-flow conditions. To increase the flexibility of water management organizations and programs and to better recognize uncertainties regarding future supplies and demands, a permitting process should be created that allows for withdrawals to be discontinued during periods of low flow and periods of comparatively high water temperature.

To reiterate and reinforce this report's six key findings and recommendations, they are repeated here:

- **Within the body of scientific literature reviewed as part of this study, the relative importance of various environmental variables on smolt survival is not clearly established. When river flows become critically low or water temperatures excessively high, however, pronounced changes in salmon migratory behavior and lower survival rates are expected. (Chapter 4).**
- **The State of Washington and other Columbia River basin entities should continue to explore prospects for water transfers and other market-based programs as alternatives**

to additional withdrawals (Chapter 6).

- The conversion of water rights to uninterruptible status will decrease the flexibility of the system during critical periods of low flows and comparatively high water temperatures. Conversions to uninterruptible rights during these critical periods are not recommended (Chapter 7).

- Sound, comprehensive Columbia River salmon management strategies will depend not only on science but also on a willingness by elected and duly appointed leaders and managers to take actions in the face of uncertainties (Chapter 7).

- Decisions regarding the issue of additional water withdrawal permits are matters of public policy, but if additional permits are issued, they should include specific conditions that allow withdrawals to be discontinued during critical periods. Allowing for additional withdrawals during the critical periods of high demand, low flows, and comparatively high water temperatures identified in this report would increase the risks of survivability to listed salmon stocks and would reduce management flexibility during these periods (Chapter 7).

- The State of Washington and other basin jurisdictions should convene a joint forum for documenting and discussing the environmental and other consequences of proposed diversions that exceed a specified threshold (Chapter 7).

References

Adams, B. L., W. S. Zaugg, and L. R. McLain. 1973. Temperature effect on parr-smolt transformation in steelhead trout (*Salmo gairdneri*) as measured by gill sodium-potassium stimulated adenosine triphosphatase. Comparative Biochemical Physiology 44A:1333-1339.

Adams, R. M., and S. H. Cho. 1998. Agriculture and endangered species protection: An analysis of tradeoffs in the Klamath Basin, Oregon. Water Resources Research 34:2741-49.

Amirfathi, P., R. Narayanan, B. Bishop, and D. Larson. 1974. A Methodology for Estimating Instream Flow Values for Recreation. Logan, UT: Utah Water Research Laboratory, Utah State University.

Anderson, J. J. 2003. Towards a resolution of the flow-survival debate. Columbia Basin Research. University of Washington.

AWWARF (American Water Works Association Research Foundation). 1998. Consumer Attitude Survey on Water Quality Issues. Denver, CO: AWWA Research Foundation.

Beeman, J., D. Rondorf, J. Faler, P. Haner, S. Sauter, and D. Venditti. 1991. Assessment of Smolt Condition for Travel Time Analysis. 1990 Annual Report to Bonneville Power Administration, DOE/BP-35245-4. Portland, OR: Bonneville Power Administration.

Beeman, J. W., D. W. Rondorf, M. E. Tilson, and D. A. Venditti. 1995. A nonlethal measure of smolt status of juvenile steelhead based on body morphology. Transactions of the American Fisheries Society 124:764-769.

Bell, M. C. 1973. Fisheries Handbook of Engineering Requirements and Biological Criteria. Portland, OR: U. S. Army Corps of Engineers, Fisheries Engineering Research Program.

Bell, M. 1991. Fisheries handbook of engineering requirements and biological criteria. Portland, OR: U.S. Army Corps of Engineers, Fish Passage Development and Evaluation Program, North Pacific Division.

Bennett, D. H., M. A. Madsen, and M. H. Karr. 1997. Water temperature characteristics of the Clearwater River, Idaho and Lower Granite, Little Goose, Lower Monumental, and Ice Harbor Reservoirs, Lower Snake River, Washington, During 1991-1993, with Emphasis on Upstream Water Releases. Data Volume II, Project 14-16-0009-1579. Moscow, ID: Idaho Department of Fish and Game.

Bentley, W. B., and H. L. Raymond. 1976. Delayed migrations of yearling Chinook salmon since completion of Lower Monumental and Little Goose dams on the Snake River. Transactions of the American Fisheries Society 105:422-424.

Berggren, T. J., and M. J. Filardo. 1993. An analysis of variables influencing the migration of juvenile salmonids in the Columbia River basin. North American Journal of Fisheries Management 13(1):48-63.

Bernardo, D. J., and N. K. Whittlesey. 1989. Factor Demand in Irrigated Agriculture Under Conditions of Restricted Water Supplies. U.S. Department of Agriculture, Economic Research Service, Technical Bulletin 1765.

Bingham, P. 2002. Forecast Summary Commodity Flow Forecast Update and Lower Columbia River Cargo Forecast. DRI-WEFA, Inc. Available online at *http://www.portofportland.com/Marine/MTMP/pdf/Presentation_Sum_CF_Frcast.pdf,* last accessed March 16, 2004.

Bjornn, T. C., and C. A. Peery. 1992. A review of literature related to movements of adult salmon and steelhead past dams and through reservoirs in the lower Snake River. Walla Walla, WA: U. S. Army Corps of Engineers, Walla Walla District.

Blumm, M. C., and B. M. Swift. 1997. A Survey of Columbia River Basin Water Law Institutions and Policies. A Report to the Western Water Policy Advisory Commission. Portland, OR: Natural Resources Law Institute Northwestern School of Law of Lewis and Clark College.

Bottom, D., C. Simenstad, A. Baptista, D. Jay, J. Burke, K. Jones, E. Casillas, and M. Schiewe. 2002. Salmon at river's end: The role of the estuary in the decline and recovery of Columbia River salmon. Seattle, WA: National Marine Fisheries Service.

BPA (Bonneville Power Administration). 1993. Seasonal Volumes and Statistics, Columbia River Basin, 1928-1989. Prepared for Bonneville Power Administration by A.G. Crook Co. Portland, OR: Bonneville Power Administration.

Brannon, E., M. Powell, T. Quinn, and A. Talbot. 2002. Population Structure of Columbia River Basin Chinook Salmon and Steelhead Trout. Final Report, Center for Salmonid and Freshwater Species at

Risk, University of Idaho, to National Science Foundation, Arlington, VA, and U.S. Department of Energy, Bonneville Power Administration, Division of Fish and Wildlife. Portland, OR: University of Idaho.

Brookshire, D. S., B. Colby, M. E. Ewers, and P. T. Ganderton. 2003. Western water markets. Submitted to Water Resources Research September 2003.

Brown, F. L., and H. Ingram. 1987. Water and Poverty in the Southwest. Tucson: University of Arizona Press.

Buettner, E. W., and A. F. Brimmer. 2000. Smolt monitoring at the head of Lower Granite reservoir and Lower Granite Dam. Annual Report 1998 Operations. Department of Energy/Bonneville Power Administration, Contract No. DE-B179-83BP11631.

Burnham, K., D. Anderson, G. White, C. Brownie, and K. Pollock. 1987. Design and analysis methods for fish survival experiments based on release-recapture. American Fisheries Society Monograph Number 5.

Butcher, W., N. Whittlesey, and J. Orsborn. 1972. Economic Value of Water in a Systems Context. NWC-SBS-72-048. Springfield, VA: National Technical Information Service.

Cada, G. F., M. D. Deacon, S. V. Mitz, and M. S. Bevelhimer. 1997. Effects of water velocity on the survival of downstream-migrating juvenile salmon and steelhead: A review with emphasis on the Columbia River. Reviews in Fishery Science 5(2):131-183.

Canby, W. C. Jr. 1981. American Indian Law in a Nutshell. St. Paul, MN: West Pub.Co.

Casavant, K. 2000. Inland Waterborne Transportation: An Industry Under Siege. Pullman, WA: USDA, Agricultural Marketing Service.

CBFWA (Columbia Basin and Fish Wildlife Authority). 2003. Memorandum to Mark Walker, Northwest Power Council. February 26.

Census of Agriculture, 1997. Agricultural Economics and Land Ownership Survey. Available online at *http://www.nass.usda.gov/census/census97/aelos/intro.htm#six*.

Chapman, D., A. Giorgi, M. Hill, A. Maule, S. McCutcheon, D. Park, W. Platts, K. Pratt, J. Seeb, L. Seeb, and F. Utter. 1991. Status of Snake River chinook salmon. Report to Pacific Northwest Utilities Conference Committee. Boise, ID: Don Chapman Consultants, Inc. Available from Don Chapman Consultants, 3653 Rickenbacker, Ste. 200, Boise, ID 83705.

Chapman, D., A. Giorgi, T. Hillman, D. Deppert, M. Erho, S. Hays, C. Peven, B. Suzumoto, and R. Klinge. 1994. Status of summer/fall

Chinook salmon in the mid-Columbia region. Available from Don Chapman Consultants, Inc., 3653 Rickenbacker, Boise, ID (now BioAnalysts, Inc., 1117 E. Plaza Dr., Suite A, Eagle, ID 83616).

Chapman, D., C. Peven, A. Giorgi, T. Hillman, and F. Utter. 1995. Status of spring chinook salmon in the mid-Columbia River. Boise, ID: Don Chapman Consultants, Inc. Available from Don Chapman Consultants, 3653 Rickenbacker, Ste. 200, Boise, ID 83705.

City of Pasco. 2003. Memorandum to Kim Holst from Doyle Heath, Utility Engineer regarding response to Email from Stu McKenzie. March 13.

Clinton, S. M., R. T. Edwards, and R. J. Naiman. 2002. Forest-river interactions: Influence on hyporheic dissolved organic carbon concentrations in a floodplain terrace. Journal of the American Water Resources Association 38:619-631.

Cohen, F. S., and R. Strickland. 1982. Felix S. Cohen's Handbook of Federal Indian Law. Charlottesville, VA: Bobbs-Merrill.

Collis, K., D. D. Roby, D. P. Craig, S. Adamany, J. Adkins, and D. E. Lyons. 2002. Colony size and diet composition of piscivorous waterbirds on the lower Columbia River: Implications for losses of juvenile salmonids to avian predation. Transactions of the American Fisheries Society 131: 537-550.

Columbia Basin Bulletin. 2002. Draft Report Details Toxic Chemicals in Basin Fish. February 15, 2002.

Confederated Tribes of the Colville Reservation. 2000. Comments on the July 27, 2000, Draft Biological Opinion on the Operation of the Federal Columbia River Power System 5.

Congleton, J. L., T. Wagner, J. Evavold, D. Fryer, and B. Sun. 2002. evaluation of Physiological Changes in Migrating Juvenile Salmonids and Effects on Performance and Survival. Annual Report to the U.S. Army Corps of Engineers DACW68-00-C-0031.

Connor, W. P., H. L. Burge, and D. H. Bennett. 1998. Detection of PIT tagged subyearling Chinook salmon at a Snake River Dam: Implications for summer flow augmentation. North American Journal of Fisheries Management 18:530-536.

Connor, W., H. Burge, and W. Miller. 1993. Rearing and emigration of naturally produced Snake River fall Chinook salmon juveniles. Pp. 86-116 In Identification of the Spawning, Rearing and Migratory Requirements of Fall Chinook Salmon in the Columbia River Basin. U.S. Fish and Wildlife Service, Lower Snake River Compensation Plan. 1991 Annual Progress Report to Bonneville Power Administration. Project No. 91-029 (Contract No. DE AI79-91BP21708).

Connor, W. P., H. L. Burge, and D. H. Bennett. 1998. Detection of subyearling Chinook salmon at a Snake River dam: Implications for summer flow augmentation. North American Journal of Fisheries Management 18:530-536.

Cooper, J. G. 1860. Explorations and Surveys to Ascertain the Most Practicable and Economical Route for a Railroad from the Mississippi River to the Pacific Ocean, 1853-5. Report to the Secretary of War. Washington, DC: Thomas H. Ford Printer.

Coues, E. 1893. The History of the Lewis and Clark Expedition. New York: Francis P. Harper.

Craig, J. A., and R. L. Hacker. 1940. The history and development of the fisheries of the Columbia River. U.S. Bureau of Fisheries Bulletin 32:133-216.

CRITFC (Columbia River Inter-Tribal Fish Commission). 1996. The Columbia River Anadromous Fish Restoration Plan of the Nez Perce, Umatilla, Warm Springs, and Yakama Tribes. Portland, OR: CRITFC.

Dauble, D. D., T. P. Hanrahan, D. R. Geist, and M. J. Parsely. 2003. Impacts of the Columbia River hydroelectric system on main-stem habitats of fall Chinook salmon. North American Journal of Fisheries Management 23: 641-659.

Dreher, K. J. 1998. Competing for the Mighty Columbia River—Past, Present and Future: The Role of Interstate Allocation. A view on Idaho's experience with flow augmentation. Presented to American Bar Association Section of Natural Resources Energy and Environmental Law. April 30-May 1. Boise, ID.

Ebbesmeyer, C. C., and W. Tangborn. 1992. Linkage of reservoir, coast, and strait dynamics, 1936-1990: Columbia River basin, Washington coast, and Juan de Fuca Strait. Pp. 288-299 In Interdisciplinary Approaches in Hydrology and Hydrogeology. St. Paul, MN: American Institute of Hydrology,

Ebel, W., and H. Raymond. 1976. Effects of atmosphere gas saturation on salmon and steelhead trout of the Snake and Columbia rivers. Marine Fisheries Review 38(7):1-14.

Faux, J., and G. M. Perry. 1999. Estimating irrigation water value using hedonic price analysis: A case study in Malheur County, Oregon. Land Economics 75(3):440-452.

FCRPS. 2001. The Columbia River System Inside Story: Federal Columbia River Power System. Second edition. Portland, OR: Bonneville Power Administration.

FPC (Fish Passage Center). 2002. Weekly Report #02-31. Portland,

OR: Fish Passage Center.

FPC. 2003. State, federal, and tribal anadromous fish manager's comments on the Northwest Power Planning Council draft mainstem amendments as they relate to flow/survival relationships for salmon and steelhead. Final document submitted to Northwest Power Planning Council. Portland , OR: Northwest Power Planning Council.

Gibbons, D. C. 1986. The Economic Value of Water. Washington, DC: John Hopkins University Press for Resources for the Future.

Giorgi, A. E., and J. W. Schlecte. 1997. Evaluation of the effectiveness of flow augmentation in the Snake River 1991-1995. DOE/BP-24576-1. Portland, OR: Bonneville Power Administration.

Giorgi, A. E., D. R., Miller, and B. P. Sanford. 1994. Migratory characteristics of juvenile ocean-type Chinook, *Oncorhynchus tshawytscha*, in John Day Reservoir on the Columbia River. Fisheries Bulletin 92:872-879.

Giorgi, A. E., T. W. Hillman, J. R. Stevenson, S. G. Hays, and C. M. Peven. 1997. Factors that influence the downstream migration rate of juvenile salmon and steelhead through the hydroelectric system in the mid-Columbia River Basin. North American Journal of Fisheries Management 17:268-282.

Giorgi, A., M. Miller, and J. Stevenson. 2002. Mainstem passage strategies in the Columbia River System: Transportation, Spill, and Flow Augmentation. Prepared for Northwest Power Planning Council. Portland, OR: Northwest Power Planning Council.

Goniea, T. M. 2002. Temperature influenced migratory behavior and use of thermal refuges by upriver bright fall Chinook salmon. M.S. thesis, University of Idaho, Moscow, ID.

Groves, P. A. 1993. Habitat available for, and used by, fall Chinook salmon within the Hell's Canyon Reach of the Snake River. 1992 Annual Progress Report, Environmental Affairs Department. Boise, ID: Idaho Power Co.

Gunderson, L. 1999. Resilience, flexibility, and adaptive management—antedotes for spurious certitude? Conservation Ecology 3(1):7. Available online at *http://www.consecol.org/vol3/iss1/art7*, last accessed March 2, 2004.

Gunderson, L. H., C. S. Holling, and S. S. Light (eds.). 1995. Barriers and Bridges to the Renewal of Ecosystems and Institutions. New York: Columbia University Press.

Gray, S. L., and R. A. Young. 1983. Economic issues in resolving conflicts in water use. Colorado Water Resources Research Institute Completion Report Series 119.

Hamlet, A. F., and D. P. Lettenmaier. 1999. Columbia River streamflow forecasting based on ENSO and PDO climate signals. Journal of Water Resources Planning and Management 125: 333-341.

Hallock, R. J., R. F. Elwell, and D. H. Fry, Jr. 1970. Migrations of adult king salmon *Oncorhynchus tshawytscha* in the San Joaquin Delta as demonstrated by the use of sonic tags. California Department of Fish and Game, Fish Bulletin 151.

Hastay, M. et al. 1971. The Columbia River as a Resource: Socioeconomic Consideration of the Diversion and the Value of Columbia River Water, Part A. State of Washington Research Center. Pullman, WA: Washington State University and the University of Washington.

Hays, S. P. 1959. Conservation and the Gospel of Efficiency: The Progressive Conservation Movement, 1890-1920. Cambridge, MA: Harvard University Press.

High, B. 2002. Effect of water temperature on adult steelhead migration behavior and survival in the Columbia River basin. M.S. thesis, University of Idaho, Moscow, ID.

Hillman, T. W., D. W. Chapman, and D. H. Bennett. 2000. Thermal effects of Potlatch effluent on Snake River fishes. BioAnalysts, Inc. Report to Potlatch Corporation. Lewiston, ID: Bionalysts, Inc.

Howe, C., and H. Ingram. 2002. Roles for the public and private sectors in water allocation: Lessons from around the world. Unpublished paper presented to Natural Resources Law Center Conference, University of Colorado Law School, June 11-14.

Howitt, R., N. Moore, and R. T. Smith. 1992. A Retrospective on California's 1991 Emerging Drought Water Bank. Report prepared for the California Department of Water Resources. Sacramento, CA: California Department of Water Resources.

Holling, C. S. (ed.). 1978. Adaptive Environmental Assessment and Management. New York: Wiley and Sons.

Houghton, J. T., Y. Ding, D. J. Griggs, M. Noguer, P. J. vander Linden, X. Dai, K. Maskell, and C. A., Johnson. 2001. Climate Change 2001: The Scientific Basis—Contribution of Working Group I to the Third Assessment Report of IPCC. Cambridge, UK: Cambridge University Press.

Huppert, D. D., and D. L. Fluharty. 1995. Economics of Snake River salmon recovery. A Report to the National Marine Fisheries Service. Research supported by Washington Sea Grant College Program, Grant Number R/MS-40. Seattle, WA.

Huppert, D., G. Green, W. Beyers, A. Subkoviak, and A. Wenzel. 2004. Economics of Columbia River Initiative. Final Report to the Washington Department of Ecology and CRI Economics Advisory Committee. Available online at *http://www.ecy.wa.gov/programs/wr/cr/Images/PDF/crieconrept_es_final.pdf,* last accessed May 26, 2004.

Instream Flow Council. 2002. Instream Flows for Riverine Resource Stewardship. The Instream Flow Council. Cheyenne, WY: Instream Flow Council.

ISAB (Independent Scientific Group). 1996. Return to the River: Restoration of salmonid fishes in the Columbia River ecosystem—development of an alternative conceptual foundation and review and synthesis of science underlying the Columbia River Basin Fish and Wildlife Program of the Northwest Power Planning Council.

ISAB (Independent Scientific Group). 2000. Return to the River 2000: Restoration of Salmonid Fishes in the Columbia River Ecosystem. NPPC 2000-12. Portland, OR: Northwest Power Planning Council.

ISAB (Independent Science Advisory Board). 2003. Review of Flow Augmentation: Update and Clarification. Northwest Planning Council.

Jackson, P. J., and A. J. Kimerling (eds.). 2003 Atlas of the Pacific North-west. Ninth edition. Corvallis, OR: Oregon State University Press.

Jay, D., and P. Naik. 2000. Climate effects on Columbia River sediment transport. Pp. 97-106 In G. Gelfenbaum and G. Kaminsky, eds, Southwest Washington Coastal Erosion Workshop Report 1999. U.S. Geological Survey Open File Report.

Jensen, I., and C. C. Shock is Jensen, l and C.C. Shock. 2002. Dealing with Drought. Malheur Agricultural Experiment Station, Ontario, OR. Available online at *http://www.cropinfo.net/Dealing with Drought.htm,* last accessed May 24, 2004.

Junge, C., and A. Oakley. 1966. Trends in production rates for upper Columbia River runs of salmon and steelhead and possible effects of changes in turbidity. Oregon Fish Comm. Research Briefs 12:22-43.

Kaczynski, V. W., and J. F. Palmisano. 1993. Oregon's Wild Salmon and Steelhead Trout: A Review of the Impact of Management and Environmental Factors. Salem, OR: Oregon Forest Industries Council.

Kane, J., and R. Osantowski. 1981. An Evaluation of Water Reuse Using Advanced Waste Treatment at a Meat Packing Plant. Proceedings of the 35th Industrial Waste Conference.

Kareiva, P., M. Marvier, and M. McClure. 2000. Recovery and management options for spring/summer Chinook salmon in the Columbia River basin. Science 290(3):977-979.

Karr, M., B. Tanovan, R. Turner, and D. Bennett. 1992. Water temperature control project, Snake River, interim report: Model studies and 1991 operations. Columbia River Inter-Tribal Fish Commission, U.S. Army Corps of Engineers, University of Idaho.

Karr, M. H., J. K. Fryer, and P. R. Mundy. 1998. Snake River water temperature control project phase II: Methods for managing and monitoring water temperatures in relation to salmon in the lower Snake River. Fisheries and Aquatic Sciences, Lake Oswego, OR.

Keenan, S. P., R. S. Krannich, and M. S. Walker. 1999. Public perceptions of water transfers and markets: Describing differences in water use communities. Society and Natural Resources 12:279-292.

Kling, G. W., K. Hayhoe, L. B. Johnson, J. J. Magnuson, S. Polasky, S. K. Robinson, B. J. Shuter, M. M. Wander, D. J. Wuebbles, and D. R. Zak. 2003. Confronting Climate Change in the Great Lakes Region. Available online at *http://www.ucsusa.org/greatlakes/pdf/confronting_climate_change_in_the_great_lakes.pdf*, last accessed May 24, 2004.

Kollar, K. L., R. Brewer, and P. H. McCauley. 1976. An Analysis of Price/Cost Sensitivity of Water Use in Selected Manufacturing Industries. Bureau of Domestic Commerce Staff Study, Water Resources Council.

Kolpin, D. W., E. T. Furlong, M. T. Meyer, E. M. Thurman, St. D. Zaugg, L. B. Barber, and H. T. Buxton. 2002. Pharmaceuticals, hormones, and other organic contaminants in U.S. streams 1999-2000: A national reconnaissance. Environmental Science and Technology 36(6):1202-1211.

Kukulka, T., and D. Jay. 2003. Impacts of Columbia River discharge on salmonid habitat: 2. Changes in shallow-water habitat. Journal of Geophysical Research 108(C9):3294.

Lee, K. N. 1989. The Columbia River basin: Experimenting with sustainability. Environment 31(6):6-11.

Lee, K. N. 1993. Compass and Gyroscope: Integrating Science and Politics for the Environment. Washington, DC: Island Press.

Lee, K. N. 1999. Appraising adaptive management. Conservation Ecology 3(2):3. Available online at *http://www.consecol.org/vol3/iss2/art3,* last accessed March 2, 2004.

Lee, K. N., and J. Lawrence. 1986. Adaptive management: Learning from the Columbia River Basin Fish and Wildlife Program. Environmental Law 16:431-433.

Lee, D. S., C. R. Gilbert, C. H. Hocutt, R. E. Jenkins, D. E. McAllister, and J. R. Stauffer, Jr. 1980. Atlas of American Freshwater Fishes. Pub. 1980-12 of the North Carolina Biological Survey. Raleigh: North Carolina State Museum of Natural History.

Leopold, L. 1994. A View of the River. Cambridge, MA: Harvard University Press.

Lin, P-C., R. M. Adams, and R. P. Berrens. 1996. Welfare effects of fishery policies: Native American treaty rights and recreational salmon fishing. Journal of Agricultural and Resource Economics 21(2):263-276.

Little, C. 2003. Redeeming the geography of hope. Natural Resources Journal 43(1):1-11.

Major, R. L., and J. L. Mighell. 1966. Influence of Rocky Reach Dam and the temperature of the Okanogan River on the upstream migration of sockeye salmon. U.S. Fish and Wildlife Service Fishery Bulletin 66:131-147.

Maloney, S. B., A. R. Tiedemann, D. A. Higgins, T. M. Quigley, and D. B. Marx. 1999. Influence of Stream Characteristics and Grazing intensity on Stream Temperatures in Eastern Oregon. General Technical Report PNW-GTR-459. Portland, OR: U.S. Department of Agriculture, Forest Service, Pacific Northwest Research Station.

McCarl, B., and M. Ross. 1985. The cost borne by electricity consumers under expanded irrigation from the Columbia River. Water Resources Research 21(9):1319-1328.

McCool, S. F., and R. W. Haynes. 1996. Projecting population growth in the interior Columbia River Basin. Research Note PNW-RN-519. U.S. Forest Service Pacific Northwest Station.

McCullough, D. A. 1999. A review and synthesis of effects of alterations to the water temperature regime on freshwater life stages of salmonids, with special reference to Chinook salmon. Columbia River Inter-tribal Fish Commission, Report for U.S. Environmental Protection Agency, Region 10. Seattle, WA: U.S. Environmental Protection Agency.

McIntosh, B. A., J. R. Sedell, J. E. Smith, R. C. Wissmar, S. E. Clarke, G. H. Reeves, and L. A. Brown. 1994. Historical changes in fish habitat for select river basins of eastern Oregon and Washington. Northwest Science 68 (Special Issue):37-53.

Meehan, W. R. (ed.). 1991. Influences of forest and rangeland management on salmonid fishes and their habitats. American Fisheries Society Special Publication 19.

Miles, E. L., A. K. Snover, A. F. Hamlet, B. Callahan, and D. Fluharty.

2000. Pacific Northwest regional assessment: The impacts of climate variability and climate change on the water resources of the Columbia River basin. Journal of the American Water Resources Association 36(2):399-420.

Miller, K. A. 2000. Managing supply variability: The use of water banks in the Western United States. Pp. 70-86 In D. A. Wilhite (ed.), Drought: A Global Assessment, Volume II. Routledge, London: Routledge.

Mohseni, O., H. G. Stefan, and J. G. Eaton. 2003. Global warming and potential changes in fish habitat in U.S. streams. Climatic Change 59:389-409.

Montgomery Water Group, Inc. 1997. Water supply, use and efficiency report. Columbia Basin Project.

Mote, P. W. 2003. Trends in snow water equivalent in the Pacific Northwest and their climate causes. Geophysical Research Letters 30(12):1601.

Mote, P., D. Canning, D. Fluharty, R. Francis, J. Franklin, A. Hamlet, M. Hershman, M. Holmberg, K. Gray-Ideker, W.S. Keeton, D. Lettenmaier, R. Leung, N Mantua, E. Miles, B. Noble, H. Parandvash, D.W. Peterson, A. Snover, and S. Willard. 1999. Impacts of Climate Variability and Change, Pacific Northwest. Seattle, WA: National Atmospheric and Oceanic Administration.

Muir, W. D., W. S. Zaugg, A. E. Giorgi, and S. McCutcheon. 1994. Accelerating smolt development and downstream movement in yearling Chinook salmon with advanced photoperiod and increased temperature. Aquaculture 123:387-399.

Naiman, R. J., T. J. Beechie, L. E. Benda, D. R. Berg, P. A. Bisson, L. H. MacDonald, M. D. O'Connor, P. L. Olson, and E. A. Steel. 1992. Fundamental elements of ecologically healthy watersheds in the Pacific Northwest coastal ecoregion. Pp. 128-188 In R. J. Naiman (ed.) Watershed Management. Balancing Sustainability and Environmental Change. New York: Springer-Verlag.

National Assessment Synthesis Team. 2000. Climate change impacts on the United States: The potential consequences of climate variability and change. Washington, DC: U.S. Global Change Research Program.

NMFS (National Marine Fisheries Service). 2000. Biological Opinion; Reinitiation of Consultation on the Operation of the Federal Columbia River Power System Including Juvenile Fish Transportation Program, and 19 Bureau of Reclamation Projects in the Columbia Basin. Available online at *http://www.nwr.noaa.gov/1hydrop/hydroweb/*

docs/Final/2000Biop.html, last accessed March 19, 2004.

Northwest Economic Associates. 2004. The Economic Benefits of Improved Water Supply Reliability in the Yakima River Basin. Report prepared for Benton County, WA.

NRC (National Research Council). 1992. Water Transfers in the West. Washington, DC: National Academy Press.

NRC (National Research Council). 1995. Science and the Endangered Species Act. Washington, DC: National Academy Press.

NRC. 1996. Understanding Risk: Informing Decision in a Democratic Society. Stern, P.D., and H. V. Fineberg (eds.). Washington, DC: National Academy Press.

NRC. 2000. Risk Analysis and Uncertainty in Flood Damage Reduction Studies. Washington, DC: National Academy Press.

NRC. 2001a. Climate Change Science: An Analysis of Some Key Questions. Washington, DC: National Academy Press.

NRC. 2001b. Aquifer Storage and Recovery in the Comprehensive Everglades Restoration Plan: A Critique of the Pilot Projects and Related Plans for ASR in the Lake Okeechobee and Western Hillsboro Areas. Washington, DC: National Academy Press.

NRC. 2002a. Riparian Areas: Functions and Strategies for Management. Washington, DC: National Academy Press.

NRC. 2002b. Privatization of Water Services in the United States: An Assessment of Issues and Experience. Washington, DC: National Academy Press.

NPCC (Northwest Power and Conservation Council). 1986. Compilation of information on salmon and steelhead losses in the Columbia River basin. Northwest Power Planning Council, Appendix D of the 1987 Columbia River Basin Fish and Wildlife Program.

NPCC. 2003. Memorandum to Council Members from Bruce Suzumoto on recent trends in fish returns to the Columbia Basin.

Olsen, D., J. Richards, and R. D. Scott. 1991. Existence values for doubling the size of Columbia River Basin salmon and steelhead runs. Rivers 2(1):44-56.

O'Neal, K. 2002. Effects of global warming on trout and salmon in U.S. streams. Washington, DC: Defenders of Wildlife.

ODFW/WDFW (Oregon Department of Fish and Wildlife/Washington Department of Fisheries and Wildlife). 1993. Status report, Columbia River Fish Runs and Fisheries. 1938-1992. Joint Report.

Ordal, E. J., and R. E. Pacha. 1963. The effects of temperature on disease in fish. Pp. 39-56 In Proceedings 12th Pacific Northwest Symposium on Water Pollution Research—Water Temperature—Influences, Effects, and Control. U.S. Public Health Service, U.S. Department of Health, Education, and Welfare. Corvallis, OR: Pacific Northwest Water Laboratory.

ODEQ (Oregon Department of Environmental Quality. 1995. Water quality standards review. Portland, OR: Department of Environmental Quality Standards and Assessments Section.

Park, D. 1969. Seasonal changes in downstream migration of age-group 0 Chinook salmon in the upper Columbia River. Transactions of the American Fisheries Society 98:315-317.

Payne, J. T., A. W. Wood, A. F. Hamlet, R. N. Palmer, and D. P Lettenmaier. 2004. Mitigating the effects of climate change on the water resources of the Columbia River basin: Climatic Change.

Pearcy, W. 1992. Ocean Ecology of North Pacific Salmonids. Seattle, WA: University of Washington Press.

Porter, C. M., and D. M. Janz. 2003. Treated municipal sewage discharge affects multiple levels of biological organization in fishes. Ecotoxicology and Environmental Safety 54: 199-206.

Prentice, E., T. Flagg, C. McCutcheon, and D. Brastow. 1990. PIT-tag monitoring systems for hydroelectric systems and fish hatcheries. Pp. 323-334 In Parker et al. Fish Marking Techniques, AFS Symposium #7.

Quigley, T. M., R. W. Haynes, R. T. Graham (eds.). 1996. Integrated Scientific Assessment for Ecosystem Management in the Interior Columbia Basin and Portions of the Klamath and Great Basins.

Quinn, T., and D. Adams. 1996. Environmental changes affecting the migratory timing of American shad and sockeye salmon. Ecology 77:1151-1162.

Raymond, H. L. 1979. Effects of dams and impoundments on migrations of juvenile Chinook salmon and steelhead from the Snake River, 1966-1975. Transactions of the American Fisheries Society 108:505-529.

Rayner, S., D. Lach, H. Ingram, and M. Houck. 2000. Weather Forecasts Are for Wimps: Why Water Resource Managers Don't Use Climate Forecasts. Final Report to NOAA Office of Global Programs. Available on-line at *http://www.ogp.noaa.gov/mpe/csi/econ-hd/fy98/rayner_final.pdf,* last accessed March 2, 2004.

Reiser, D. W., and T. C. Bjornn. 1979a. Habitat requirements of anadromous salmonids. Meehan W. R. Influence of forest and rangeland management on anadromous fish habitat in the western United States and Canada. PNW-GTR-96. Portland, OR: U.S. Forest Service.

Reiser, D. W., and T. C. Bjornn. 1979b. Influence of forest and rangeland management on anadromous fish habitat in western North America: Habitat requirements of anadromous salmonids. PNW-GTR-96. Portland, OR: U. S. Forest Service.

Rieman, B. E., R. C. Beamesderfer, S. C. Vigg, and T. P. Poe. 1991. Estimated loss of juvenile salmonids to predation by northern squawfish, walleyes, and smallmouth bass in John Day Reservoir, Columbia River. Transactions of the American Fisheries Society 120:448-458.

Riebsame, W. (ed.). 1997. Atlas of the New West. New York: W. W. Norton and Company.

Robards, M., and T. Quinn. 2002. The migratory timing of adult summer-run steelhead in the Columbia River over six decades of environmental change. Transactions of the American Fisheries Society 131:523-536.

Rugerrone, G. 1986. Consumption of migrating juvenile salmonids by gulls foraging below a Columbia River dam. Transactions of the American Fisheries Society 115:736-742.

Sauter, S. T., J. McMillan, and J. Dunham. 2001. Salmonid behavior and water temperature. U.S. Environmental Protection Agency, Issue Paper 1, prepared as part of EPA Region 10 Temperature. The Water Quality Criteria Guidance Development Project, EPA-910-D-01-001.

Schindler, D. E., M. D. Scheuerell, J. W. Moore, S. M. Gende, T. B. Francis, and W. J. Palen. 2003. Pacific salmon and the ecology of coastal ecosystems. Frontiers in Ecology and the Environment 1(1):31-37.

Schultz, I. R., A. D. Skillman, J. Nicolas, D. G. Cyr, and J. J. Nagler. 2003. Short term exposure to 17alpha-ethynylestradiol decreases the fertility of sexually maturing male rainbow trout (*Oncorhynchus mykiss*). Environmental Toxicology and Chemistry 22(6):1272-1280.

Service, R. P. 2004. As the West Goes Dry. Science 303:1124-1227.

Shelton, B. L. 1997. A View from Front Lines: Current Status of Four Water Right Cases. Available online at *http://www.ucowr.siu.edu/updates/pdf/V107_A3.pdf*, last accessed May 24, 2004.

Sherwood, C., D. Jay, R. Harvey, P. Hamilton, and C. Simenstad. 1990. Historical changes in the Columbia River estuary. Progressive Oceanography 25:299-352.

Simenstad, C., K. Fresh, and E. Salo. 1982. The role of Puget Sound and Washington coastal estuaries in the life history of Pacific salmon: An unappreciated function. Pp. 343-364 In V. Kennedy (ed.), Estuarine Comparisons. New York: Academic Press.

Sims, C. W., and F. Ossiander. 1981. Migrations of juvenile Chinook salmon and steelhead in the Snake River, from 1973 to 1979, a research summary. Report to U. S. Army Corps of Engineers, Portland, OR. Prepared by National Marine Fisheries Service, Contract No. DACW68-78-C-0038.

Sims, C.W., R.C. Johnsen, and W. W. Bentley. 1976. Effects of power peaking operations on juvenile salmon and steelhead trout migrations 1975. Seattle, WA: National Marine Fisheries Service, Northwest Fisheries Center.

Sims, C.W., W. W. Bentley, and R.C. Johnsen. 1977. Effects of power peaking operations on juvenile salmon and steelhead trout migrations—Progress 1976. NOAA-NMFS-CZES, NW/Alaska Fisheries Center. Contract No. DACW68-77-C-0025. Seattle WA: U.S. Army Corps of Engineers.

Sims, C.W., W. W. Bentley, and R.C. Johnsen. 1978. Effects of power peaking operations on juvenile salmon and steelhead trout migrations—Progress 1977. NOAA-NMFS-CZES, NW/Alaska Fisheries Center. Contract No. DACW68-77-C-0025. Seattle WA: U.S. Army Corps of Engineers.

Smith, S. G., W. D. Muir, R. W. Zabel, E. E. Hockersmith, G. Axel, W. Connor, and B. Arnsberg. 2002. Survival of hatchery subyearling fall Chinook salmon in the free-flowing Snake River and lower Snake River reservoirs, 1998-2001. Report of research submitted to Bonneville Power Administration by National Marine Fisheries Service.

Smith, S. G., W. D. Muir, E. E. Hockersmith, R. W. Zabel, R. J. Graves, C. V. Ross, W. P. Connor, and B. D. Arnsberg. 2003. Influence of river conditions on survival and travel time of Snake River subyearling fall Chinook salmon. North American Journal of Fisheries Management 23:939-961.

Sommer, T., M. Nobriga, W. Harrell, W. Batham, and W. Kimmerer. 2001. Floodplain rearing of juvenile Chinook salmon: evidence of enhanced growth and survival. Canadian Journal of Fisheries and Aquatic Science 58:325-333.

Tarlock. A. D. 2000. The Law of Water Rights and Resources. New York: Clark Boardman and Co.

U.S. Bureau of Reclamation. 2002. Interim Operating Plan for the Yakima River Basin.

USGS (U.S. Geological Survey). 1996. Water Resources Data, Oregon,

Water Year 1995. U.S. Geological Survey Water-Data Report OR-95-1,452.

USGS. 1998. Water quality in the Central Columbia plateau, Washington and Idaho, 1992-95. USGS Circular 1144.

USGS. 2003a. Annual volume pumped at Grand Coulee Dam. Available online at *http://nwis.waterdata.usgs.gov/wa/nwis/discharge,* last accessed March 17, 2004..

USGS. 2003b. Central Columbia River Plateau-Yakima River Basin CCYK, NAWQA Cycle II Study Area. Available online at *http://wa.water.usgs.gov/ccyk/*, last accessed March 17, 2004.

U.S. National Assessment. 2000. Chapter 3 Table 3.4 Climate Change Impacts on the United States The Potential Consequences of Climate Variability and Change Overview: Pacific Northwest By the National Assessment Synthesis Team, US Global Change Research Program.

Vigg, S., and C. C. Burley. 1991. Temperature-dependent maximum daily consumption of juvenile salmonids by northern squawfish (*Ptychocheilus oregonensis*) from the Columbia River. Canadian Journal of Fisheries and Aquatic Sciences 48:2491-2498.

Volkman, J. M. 1997. A River in Common: The Columbia River, the Salmon Ecosystem, and Water Policy. A Report to the Wester Water Policy Review Advisory Commission. Springfield, VA: National Technical Information Service.

Volkman, J. M., and W. McConnaha. 1993. Through a glass, darkly: Columbia River salmon, the Endangered Species Act, and adaptive management. Environmental Law 23(4): 1249-1272.

Washington Department of Ecology. 2003a. River and stream water quality monitoring. Available online at *http://www.ecy.wa.gov/apps/watersheds/riv/station.asp,* last accessed on March 17, 2004..

Washington Department of Ecology. 2003b. Long term river and stream water quality monitoring stations: 411A079.xls crb ck nr Beverly.asp, 36A070.xls colr at vernita.asp, 53A070.xls col r gd coulee.asp. Available online at *http://www.ecy.wa.gov/apps/watersheds/riv/station.asp,* last accessed March 17, 2004..

Waldeck, D. A., and E. H. Buck. 1999. The Pacific Salmon Treaty: The 1999 Agreement in Historical Perspective. Congressional Research Service Report RL30234.

Walsh, R. G., R. Aukerman, and R. Milton. 1980. Measuring Benefits and the Economic Value of Water in Recreation on High Country Reservoirs. Colorado Water Resources Research Institute Completion Report No. 102. Fort Collins, CO: Colorado State University.

Walters, C. J. 1986. Adaptive Management of Renewable Resources. New York: MacMillan .

WDFW-ODFW (Washington Department of Fish and Wildlife and the Oregon Department of Fish and the Oregon Department of Fish and Wildlife). 2002. Status Report: Columbia River Fish Runs and Fisheries, 1938-2000. Available online at *http://wdfw.wa.gov/fish/columbia/2000_status_report_text.pdf,* last accessed February 17, 2004.

Weber, K. R. 1990. Effects of water transfers on rural areas: A response to Shupe, Weatherford and Checchio. Natural Resources Journal 30(Winter):13-15.

White, R. 1996. The Organic Machine: The Remaking of the Columbia River. New York: Hill and Wong.

Williams, J. G., and G. M. Matthews. 1995. A review of flow and survival relationships for spring and summer Chinook salmon, *Oncorhynchus tshawytscha*, from the Snake River Basin. Fisheries Bulletin 93:732-740.

Wissmar, R. C., J. E. Smith, B. A. McIntosh, H. W. Li, G. H. Reeves, and J. R. Sedell. 1994. A history of resource use and disturbance in riverine basins of eastern Oregon and Washington (early 1800s-1990s). Northwest Science 68:1-35.

Young, F. R., 1997a. Development of a systemwide predator control program: Stepwise implementation of a predation index, predator control fisheries and evaluation plan in the Columbia River Basin (Northern squawfish management program): Section 2—evaluation. DOE/BP-24514-4. Portland, OR: Bonneville Power Administration.

Young, F. R., 1997b. Development of a systemwide predator control program: stepwise implementation of a predation index, predator control fisheries and evaluation plan in the Columbia River basin (northern squawfish management program) Section 1: Implementation. DOE/BP-24514-3. Portland, OR: Bonneville Power Administration.

Zabel, R. W., S. G. Smith, W. D. Muir, D. M. Marsh, J. G. Williams, and J. R. Skalski. 2002. Survival estimates for the passage of spring-migrating juvenile salmonids through Snake and Columbia river dams and reservoirs, 2001. Prepared for Department of Energy/Bonneville Power Administration, Contract No.DE-AI79-93BP10891, Project No. 93-29.

Zaugg, W. S. 1987. Smolt indices and migration. Pp. 50-60 In Improving Hatchery Effectiveness as Related to Smoltification. Proceedings of a Workshop May 20-23, 1985, Kah-Nee-Tah Lodge, Warm Springs, OR.

Zaugg, W. S., and H. Wagner. 1973. Gill ATPase activity related to parr-smolt transformation and migration in steelhead trout (*Salmo gairdneri*): Influence of photoperiod and temperature. Comparative Biochemisty and Physiology 45:955-965.

Zaugg, W. S., B. L. Adams, and L. R. McLain. 1972. Steelhead migration: Potential temperature effects as indicated by gill adensosine triphosphatase activities. Science 176:415-416.

Zimmerman, M. 1999. An overview of Columbia River predation studies. Pp. 89-92 in ODFW and NMFS. 1999. Management Implications of Co-occurring Native and Introduced Fishes: Proceedings of the Workshop. October 27-28, 1998, Portland, OR.

Appendixes

Appendix A

Columbia River Initiative Draft Management Scenarios
July 7, 2003

BACKGROUND

The Washington State Department of Ecology has developed the following set of alternative draft management scenarios as the next step in the Columbia River Initiative (CRI). The draft management scenarios reflect a range of potential water resources management strategies for the Columbia River mainstem. Each scenario describes a specific hypothetical management approach to water use and mitigation, if required, and generally describes the approach that would be used by Department of Ecology decisionmakers as they review water rights applications.

The scope of work for the National Research Council's committee includes a requirement to review and comment on a set of management scenarios to be provided by the Department of Ecology. In the form described herein, the alternative scenarios represent early thinking about a range of possible outcomes relating risk to salmon and water use and establishing sufficient difference for scientific consideration. They should not be interpreted as a set of final proposals, nor as a package intended to constrain the potential outcomes of the scientific review. The management program that is eventually proposed by the Department of Ecology as a formal rule will have been shaped by feedback from the scientific review and would likely include elements that are yet to be suggested by interested parties.

As information becomes available from the science review, a management program will be developed for further refinement and will be drafted as a proposed rule by the Department of Ecology. Both formal and informal public review and comment will be included as elements of the rule-making process. Final adoption of the rule will take place following publication of the National Research Council's report.

The management program developed as the basis for rule making will become the most important product to result for the CRI. The guidance this program will provide to the Department of Ecology would in large part define the permitting program with regard to new water allocations and mitigation decisions and would be the basis on which the State of Washington implements its dual responsibilities to manage water resources and protect the environment.

FIVE MANAGEMENT SCENARIOS

The following five draft scenarios are submitted to the National Research Council for review. With the exception of the No Action Scenario, each scenario describes an amount of water to be allocated for out-of-stream use and any mitigation that might be undertaken in conjunction with the increased use of water. The scenarios are further distinguished based on a set of premises regarding the risk to salmonid populations that would arise from additional water withdrawals from the mainstem of the Columbia River.

Scenario 1: Water Allocation Linked to Current Salmon Efforts

The key premise of Scenario 1 is that there is a low risk to salmon survival resulting from existing and new allocations of water and that the state's current salmon recovery efforts are adequate, that is, the benefits from current efforts exceed the risks associated with new water allocations. For Scenario 1 it is assumed that the state and region will continue to make current or increased investments in existing salmon recovery-related environmental activities but that these investments are relatively unrelated to a new Washington water resources management program that would allocate or recognize up to 2,000,000 acre-feet of new water over a 20-year period, 1,000,000 of which would be for out-of-stream uses in Washington State.

As embodied in the Northwest Power Planning Council's Fish and Wildlife Plan and Washington's Statewide Strategy to Recover Salmon, existing salmon-related environmental activities include direct investments in salmon recovery projects made by the Salmon Recovery Funding Board and local salmon recovery groups, state and local investments in watershed planning, ongoing efforts to establish instream flows in tributaries to the Columbia River, the state program to purchase water rights to support instream flows, and state and federal funding of irrigation efficiency. (Detailed descriptions of these programs will be provided to the National Research Council committee.)

In Scenario 1 it is assumed that water resources could be made available for use between the Canadian border and Bonneville Dam. New permits would be issued by the State of Washington during a 20-year window, not to exceed 1,000,000 acre-feet in total. Within the total amount of water allocated by Scenario 1, approximately 220,000 acre-feet would be made available to meet demand within the Columbia Basin Project. In addition to the 1,000,000 acre-feet to be allocated to Washington water users by Scenario 1,427,000 acre-feet, representing flow and temperature management actions taken in the Snake River, would be legally recognized through the Washington State reaches of the Snake and Columbia rivers, and 600,000 acre-feet would be recognized as necessary to meet the water resources needs of the state of Oregon. Commitments of water resources in this scenario total 2,000,000 acre-feet, of which 1,600,000 could be developed for out-of-stream use over the next 20 years.

Permits that are currently subject to interruption when streamflows reach a predetermined level could, at the owner's option, be converted to uninterruptible status. These water rights could be converted to uninterruptible status by demonstrating that current water use conforms to state-of-the-art water use efficiency practices. Likewise, all new water rights issued by the state would require state-of-the-art efficiency in proposed uses and would also be metered.

Periodic assessment of the state's water resources management program would be integral and ongoing. Scientific information would be used to adapt the program as necessary to accommodate changes in knowledge over time. Formal reevaluations of the program would take place at year 10 and year 20.

In addition, the state would seek partners to create a functioning water market or "water bank" for the mainstem of the Columbia River to facilitate a more efficient allocation of existing water resources in the basin.

Scenario 2: Incremental Mitigation Linked to New and Modified Permits

Scenario 2 presumes that a new level of contribution to salmon health and recovery would be required to secure sufficient additional benefits for fish and to offset the risk created by additional water withdrawals from the river. Revenue to support the additional level of effort would be generated by a $10 per acre-foot per year usage charge on new permits and on existing rights that are converted from an interruptible to an uninterruptible status. The elements of the scenario would be in addition to the ongoing state and regional actions, assessment, and water bank described in Scenario 1.

New permits would be issued during a 20 year window, not to exceed 700,000 acre-feet in total. The state would issue an additional 300,000 acre-feet (a total of 1,000,000 acre-feet) from the mainstem once existing users demonstrate that conservation investments are in place for a majority of water users on the mainstem. Applicants for new permits or conversion of existing permits to uninterruptible status would also be required to demonstrate compliance with state-of-the-art efficiency standards.

Revenue generated would provide funds to acquire mitigation water in low-water years and to make habitat improvements in the mainstem and tributaries. In addition to existing salmon-related environmental activities, the development of storage projects could be explored using these resources. Fishery managers would be asked to prioritize the use of these resources and would consider implementing a low water year strategy.

Scenario 3: Enhanced Level of Mitigation

This alternative would incorporate the current salmon recovery-related environmental activities and other proposed actions described in Scenarios 1 and 2. However, this scenario is premised on the notion that a more robust contribution to salmon health and survival would be necessary to secure additional benefits to fish and to offset the risks caused by additional water withdrawals from the river. Revenue to support the additional level of effort would be generated by a $20 per acre-foot per year usage charge on new permits and on existing rights that are converted from an interruptible to an uninterruptible status. Revenue generated by the usage charge would be used to benefit salmon recovery projects. Consistent with Scenario 2, this alternative would create a 20-year window to issue new water use permits, in an amount not to exceed 1,000,000 acre-feet in total.

To supplement actions supported by the usage charge on new permits and on existing rights that are converted to an uninterruptible status, the state would provide financial support to install new conservation measures. The state would also actively explore other means to provide additional water for offstream and instream uses (e.g. storage developments). Fishery managers would be asked to prioritize the use of these resources, and would consider implementing a low water year strategy.

Scenario 4: In-Place, In-Kind, and In-Time Mitigation

Scenario 4 assumes that the risk to salmonid survival that would result from additional water withdrawals from the Columbia River is so significant that it must be directly offset in proportion to consumption. No new water rights would be permitted without being offset by direct mitigation in the mainstem of the Columbia River.

Under Scenario 4, all new water rights could be required to offset water use through water rights changes and transfers, conservation, and/or utilizing newly developed storage capacity. The state would pursue conservation savings from existing rights and would also actively pursue storage projects that could provide the capacity to support new water resources for out-of-stream appropriation.

Existing water rights could be converted to an uninterruptible status by conforming to state-of-the-art water use efficiency standards and by paying a $30 per acre-foot per year usage charge. Revenue generated would provide funds to acquire mitigation water in low water years and to make habitat improvements in the mainstem and tributaries.

Scenario 5: No Action Scenario

Scenario 5 assumes that the existing rule governing the water resources of the Columbia River, the Department of Ecology would require consultation with fish managers (Washington Department of Fish and Wildlife, Tribes, National Oceanic and Atmospheric Administration—Fisheries Division) prior to allocating new water rights. Under this scenario whether or not mitigation is required and the type and quantity of that mitigation are decisions made on each permit on a case-by-case base as a result of the consultation.

Appendix B

Resources Group

1. James Anderson, University of Washington, Seattle

2. Hal Beecher, Washington Department of Fish and Wildlife, Olympia

3. John Covert, Washington Department of Ecology, Olympia

4. Steve Hays and Joe Lukas, Mid-Columbia Public Utilities Districts

5. Robert Heineth, Columbia River Inter-Tribal Fish Commission, Portland, Oregon

6. Nate Mantua, University of Washington, Department of Atmospheric Sciences, Seattle

7. Tony Nigro, Oregon Department of Fish and Wildlife, Salem

8. Charley Petroskey, Idaho Department of Fish and Game, Boise

9. Howard Schaller, U.S. Fish and Wildlife Service, Vancouver

10. Paul Wagner, NOAA Fisheries, Seattle

Appendix C

Calculations on Annual Discharges of Water from the Columbia Basin Project

Gauged data are available on water withdrawals from Lake Roosevelt that serve as the principal supply water to the three irrigation districts in the Columbia Basin Project (CBP). In contrast, the total discharge from CBP that returns back into the main stem of the Columbia River is not measured or estimated. An attempt is hereby made to calculate irrigation return flows through an annual mass (volume) balance on water (Tanji and Kielen, 2002). The annual mass balance on water from an irrigation project is defined as:

Volume Water Inflows − Volume Water Outflows
$= \pm \Delta$ *Storage* (1)

If the control volume (system of interest) for CBP includes both the vadose and saturated zones of the CBP, Eq. (1) expands to

(Surface Water Inflows + Subsurface Water Inflows) − (Surface Water Outflows + Subsurface Water Outflows)
$= \pm \Delta$ *Storage* (2)

For a comprehensive mass balance on water in the CBP, the components of inflows and outflows may include:

Surface Water Inflows = Irrigation Water + Precipitation + Captured natural rim inflows (3)

Subsurface Water Inflows = Groundwater Rim Inflows + Seepage Inflow from River (4)

Surface Water Outflows = Crop ET + Non-crop ET + Precipitation E&ET + Reservoir evaporation + Irrigation Canal & Lateral Evaporation + Drain Canal Evaporation + Operational and Lateral Spills + Surface Irrigation Drainage into River (5)

Subsurface Water Outflows = Groundwater Rim Outflows + Groundwater Outflows into River + Phreatophyte ET (6)

Natural rim inflows refer to surface water inflows from the watershed into the CBP, such as Crab Creek watershed that is impounded in the Potholes Reservoir for use as irrigation water. Groundwater rim inflows are the subsurface inflows of groundwater from lands adjacent to the CBP. Seepage inflows from the river denote subsurface inflows into the CBP from the main stem of the Columbia River. The symbol *ET* is defined as evaporation losses (*E*) from moist soil and transpiration (*T*) losses of water from cropped plants as well as noncropped or native vegetation other than phreatophytes that extract water from the saturated zone such as open drains and wetlands. Groundwater rim outflows are subsurface flows from the CBP to adjacent lands, and groundwater outflows into river are subsurface accretions of water into the Columbia River. The above components of water flows are typically available only when an irrigation project has been subjected to detailed hydrological investigations and/or hydrological modeling.

Over decades, $\pm\Delta$ Storage in Eq. (1) may be assumed to be zero, so that

Water Inflows = Water Outflows (7)

The irrigation return flow (IRF) from the CBP into the Columbia River consists of spills from canals and laterals, surface irrigation drainage and groundwater outflow into the river. When data such as surface irrigation drainage and subsurface outflows into the river are not available, (as in the case at the CBP), the above mass balance equations may be used to obtain these flows as a closure term (i.e., by difference). For the case of the CBP, the principal missing data are surface irrigation drainage for surface water outflows into the river as well as groundwater outflow into river for subsurface water outflows.

Fortunately, the *Water Supply, Use and Efficiency Report* regarding the Columbia Basin Project is available from Montgomery Water Group, Inc. (1997). This report, however, does not contain all the water flow components identified in Eqs. (3) to (6), and therefore a more simplified water balance is utilized taking into consideration only the major components of water flow. The rationale for the simplification and the neglect of certain flow components is as follows:

1. Annual average precipitation in the CBP is only about 10.1 inches, much of which is lost through ET, and hence precipitation and precipitation E&ET may be neglected.
2. Groundwater rim inflows into the CBP and rim outflow from the CBP as well as seepage from the Columbia River into the CBP are difficult components to estimate and herein assumed to cancel each other.
3. Noncrop ET or ET from native vegetation is assumed to be small as compared to crop ET and, because of low annual precipitation, ET from phreatophytes is also assumed to be small.

If one accepts the above assumptions and simplifications, the annual mass balance on water in the CBP may be rearranged to

(*River Withdrawal* + *Captured Natural Rim Inflows*) − (*Crop ET* + *Reservoir Evaporation* + *Canal & Lateral Evaporation* + *Operational and Lateral Spills*) = (*Surface Irrigation Drainage* + *Groundwater Outflows into River*) (8)

Appendix Table C-1 contains the annual mass balance on water for the CBP from 1975 through 1994. Column J gives the combined surface irrigation drainage and groundwater outflow into the Columbia River, the closure term. In this mass balance it is not possible to separate out groundwater outflow from surface irrigation drainage. The latter could be monitored comparatively easily but not the former. The ratio of irrigation return flow to total inflow averages 0.30 (30 percent) of supply water. This also means consumptive water use (evaporated to the atmosphere) is 70 percent because ΔS is assumed to be zero.

The irrigation return flow ratio for the CBP is similar to those of irrigation districts in California, for example, Glenn-Colusa Irrigation District in the Sacramento Valley (0.29) and Panoche Water District in the San Jaoquin Valley (0.31) (Tanji, 1981) and Imperial Irrigation District in the Imperial Valley (0.36) (Kaddah and Rhoades, 1976).

References

Kaddah, M. T., and J. D. Rhoades. 1976. Salt and water balance in Imperial Valley, California: Soil Science Society of America Journal 49:93-100.

Montgomery Water Group, Inc. 1997. Water supply, use and efficiency report, Columbia Basin Project: 47 pp. plus figures and tables.

Tanji, K. K. 1981. California irrigation return flow case studies: ASCE Journal of Irrigation and Drainage Division 107(IR2):209-220.

Tanji, K. K., and N. C. Kielen. 2002. Agricultural drainage water management in arid and semi-arid climates. FAO Irrigation and Drainage Paper 61, Food and Agriculture Organization of the United Nations, Rome, Italy.

234 Managing the Columbia River

	A	B	C	D	E	F	G	H	I	J	K	L
1	Table Mass balance on water for Columbia Basin Project in millions of ac-ft/yr											
2	Closure on unmeasured flows:											
3	(Irrig water + Natural inflow) - (Crop ET + Res&Canal evap + Canal&Lateral spills) = (Surface irrig drainage + Ground water outflow to river)											
4												
5	A	B	C	D	E	F	G	H	I	J	K	L
6				(B+C)					(E+F+G+H)	(D-I)	(H+J)	(K/D)
7				Total			Canal&Lat	Canal&Lat				
8	Year	Irrig water	Nat inflow	inflow	CropET	ResEvap	Evap	Spills	Losses	Closure	IRF	Ratio
9	1975	2.14	0.271	2.411	1.32	0.179	0.182	0.191	1.872	0.539	0.73	0.303
10	1976	2.34	0.24	2.58	1.349	0.185	0.218	0.221	1.973	0.607	0.828	0.321
11	1977	2.627	0.057	2.684	1.4	0.181	0.219	0.21	2.01	0.674	0.884	0.33
12	1978	2.247	0.17	2.417	1.386	0.179	0.213	0.234	2.012	0.405	0.639	0.264
13	1979	2.671	0.117	2.788	1.309	0.182	0.238	0.203	1.932	0.856	1.059	0.38
14	1980*	1.454	0.139	1.593	1.315	0.176	0.199	0.241	1.931	-0.338	-0.097	-0.06
15	1981	2.913	0.079	2.992	1.303	0.177	0.213	0.245	1.938	1.054	1.299	0.434
16	1982	2.394	0.146	2.54	1.306	0.176	0.205	0.25	1.937	0.603	0.853	0.336
17	1983	2.055	0.156	2.211	1.355	0.17	0.202	0.249	1.976	0.235	0.484	0.219
18	1984	2.244	0.286	2.53	1.33	0.174	0.212	0.285	2.001	0.529	0.814	0.322
19	1985	2.269	0.118	2.387	1.37	0.177	0.226	0.271	2.044	0.343	0.614	0.257
20	1986	2.641	0.171	2.812	1.353	0.177	0.231	0.278	2.039	0.773	1.051	0.374
21	1987	2.548	0.117	2.665	1.324	0.176	0.267	0.29	2.057	0.608	0.898	0.337
22	1988	2.741	0.063	2.804	1.352	0.183	0.256	0.342	2.133	0.671	1.013	0.361
23	1989	2.621	0.168	2.789	1.352	0.179	0.25	0.294	2.075	0.714	1.008	0.361
24	1990	2.712	0.04	2.752	1.375	0.179	0.255	0.31	2.119	0.633	0.943	0.343
25	1991	2.773	0.071	2.844	1.393	0.178	0.277	0.309	2.157	0.687	0.996	0.35
26	1992	2.729	0.05	2.779	1.393	0.174	0.264	0.29	2.121	0.658	0.948	0.341
27	1993	2.417	0.102	2.519	1.482	0.179	0.225	0.302	2.188	0.331	0.633	0.251
28	1994	2.354	0.058	2.412	1.463	0.179	0.25	0.287	2.179	0.233	0.52	0.216
29	ave	2.444	0.131	2.575	1.362	0.178	0.23	0.265	2.035	0.54	0.805	0.302
30												
31	*Eruption of Mount St Helens interrupted irrigation											

Appendix D

Climate Change and Hydrological Impacts

The regional climate of the Pacific Northwest influences water temperatures, the flows of the Columbia River, and soil moisture and groundwater availability in the Columbia basin. The flows and temperature requirements for salmonids resources and threatened and endangered stocks should be evaluated in the context of historical and potential future variability and change in both water temperatures and streamflow. Prospective changes in climate are important, as climate shifts over the past 30 years have produced shifts in the distributions and abundance of many species and appear to be responsible for one species-level extinction (Thomas et al., 2004).

The regional climate influences water temperatures of the Columbia River basin. These water temperatures have been increasing over the last 45 years (1953 to 1998) in the Columbia River at a rate of about 0.38°C per decade or 1.9°C per 50 years (Figure 3-8). Some of this increase can arguably be accounted for by nonclimatic changes in the river basin such as dams and reservoirs, changes in land use, increases in water withdrawals, and other factors. However, the nearest river to the Columbia River of similar dimensions is the undammed Fraser River in Canada, which also has experienced temperature increases from 1953 to 1998 of about 0.2°C per decade or almost 1°C per 50 years (British Columbia Ministry of Water, Land and Air Protection). Average August temperatures of the Columbia River (Figure 3-8) are now about 5°C higher than the average summer temperatures of the Fraser River.

Historically, winter conditions contributing to winter snowpack, maximum streamflow in spring, and maintenance of summer and even winter flows have varied greatly over the last century. They are expected to vary and change in the future.

The influence of interyear and interdecadal variability on the hydrograph at the Dalles Dam from 1900 to 1998 have been summarized by Miles et al. (2000; see also Hamlet and Lettenmaier, 1999). A dominant source of the inter-year variability in flows has been driven by the climate variability associated with El Niño-Southern Oscillation (ENSO) and La Niña conditions. The Pacific Decadal Oscillation (PDO) also drives variability of flows (Miles et al., 2000). These two large-scale climatic drivers (ENSO and PDO) can interact to affect the lowest and the highest flows. Although these climate change drivers are important and must be noted, a detailed analysis of them was beyond the scope of this report.

Prospective future climate changes (driven by greenhouse gas emissions) have been simulated, with many simulation model results suggesting that the water supply of the Columbia River may be reduced in the next half century. Scenarios of future changes in the Columbia River hydrograph suggest that future warming will move the river toward conditions, on average, that closely resemble conditions observed during the warm phases of ENSO and PDO during the last century (Hamlet and Lettermaier, 1999; Miles et al., 2000). These simulations were generated with two general circulation models (GCMs) for the years 2025, 2045, and 2095 using expected rates of carbon dioxide emissions. One model was from the Max Plank Institute in Germany and the other was the Hadley 2 model from the Hadley Center in the United Kingdom. Both models indicate warming in all months relative to historical air temperature from 1961 to 1997. For 2045 the projected air temperature increases in individual months range from about 1° to about 4°C. The fact that the Hadley 2 model projects wetter conditions than observed historically especially in summer and fall, while the Max Planck model projects dryer conditions in the summer and fall, demonstrates the uncertainties associated with climate change model projections of changes in precipitation associated with temperature increases. As noted, the models are more consistent in projecting temperature increases.

References

British Columbia Ministry of Water, Land and Air Protection. Summaries in part by John Morrison, Institute of Ocean Sciences. Web materials from the Government of British Columbia at *http://britishcolumbia.c-ciarn.ca/*.

Hamlet, A. F., and D. P. Lettenmaier. 1999. Effects of climate change on hydrology and water resources in the Columbia basin. Journal of the American Water Resources Association. 35(6):1597-1623.

Miles, E. L., A. K. Snover, A. F. Hamlet, B. Callahan, and D. Fluharty. 2000. Pacific Northwest Regional assessment: The impacts of climate variability and climate change on the water resources of the Columbia River basin. Journal of the American Water Resources Association 36:399-420.

Thomas, C. D., A. Cameron, R. E. Green, M. Bakkenes, L. J. Beaumont, Y. C. Collingham, B. F. N. Erasmus, M. F. de Siquiera, A. Grainger, L. Hannah, L. Hughes, B. Huntley, A. S. van Jaarsveld, G. F. Midgley, L. Miles, M. A. Ortega-Huerta, A. T. Peterson, O. L. Phillips, and S. E. Williams. 2004. Extinction risk from climate change. Nature 427(8):145-148.

Appendix E

Committee Biographical Information

ERNEST T. SMERDON, *Chair*, recently retired as vice-provost and dean of the College of Engineering and Mines at the University of Arizona. Dr. Smerdon has served as an advisor to the U.S. federal government and several foreign governments on water resources and agricultural development issues for four decades. He has authored over 100 professional papers on water resources planning, engineering, and irrigation. He has also served on several National Research Council committees and boards. He is a member of the National Academy of Engineering. Dr. Smerdon received his B.S. degree, his M.S. degree, and his Ph.D. degree, all in engineering, from the University of Missouri, Columbia.

RICHARD M. ADAMS is a professor of agricultural and resource economics at Oregon State University. Prior service includes assistant and associate professor, University of Wyoming. He has served as editor of the *American Journal of Agricultural Economics* and associate editor for *Water Resources Research and the Journal of Environmental Economics and Management*. He is a member of various government committees dealing with climate change and air and water pollution. His current research focuses on the economic effects of air and water pollution, implications of climate change for agriculture and water resources, and tradeoffs between agricultural activity and environmental quality. Dr. Adams received his B.S. degree in resource management, his M.S. degree in agricultural economics, and his Ph.D. degree from the University of California, Davis.

DONALD W. CHAPMAN is a consulting biologist who lives in Eagle, Idaho. He was an inland fishery and stock assessment biologist with U.N. FAO in Cartagena, Colombia, and Kigoma, Tanzania. Earlier he was a professor and fishery unit leader at the University of Idaho and a visiting professor at Montana State University and the University of Wisconsin. He formerly was director of research for the Oregon Fish Commission, executive secretary of the Oregon State Water Resources Research Institute, and coordinator of the Alsea Watershed Study. His research interests include catch and stock assessment, anadromous fish passage problems, habitat evaluations, salmonid ecology, and fishery resource management. He received his B.S. degree in forest management and his M.S. and Ph.D. degrees in fisheries from Oregon State University.

DARRELL G. FONTANE is director of the International School for Water Resources and a professor in the civil engineering department at Colorado State University. His research interests include water resources decision support systems, computer-aided water management, and integrated water quantity and quality management. As director of the International School for Water Resources, he organizes nondegree programs for international engineers in various aspects of water resources engineering. His responsibilities also include graduate teaching and research in water resources systems analysis and computerized decision support systems. Dr. Fontane received his B.S. degree in civil engineering from Louisiana State University, his M.S. degree in civil engineering from the Georgia Institute of Technology, and his Ph.D. degree in civil engineering from Colorado State University.

ALBERT E. GIORGI is president and senior fisheries scientist at BioAnalysts, Inc., in Redmond, Washington. Dr. Giorgi has been conducting research on Pacific Northwest salmonid resources since 1982. Previously, he conducted research in marine invertebrate ecology and marine fish life history. In his capacity as a salmon biologist he specializes in migratory behavior, juvenile salmon survival studies, biological effects of hydroelectric development and operation, and population modeling. Dr. Giorgi's clients include the Bonneville Power Administration; Northwest Power Planning Council; Corps of Engineers; Chelan, Douglas, and Grant County Public Utility Districts; and several engineering firms including CH_2M Hill, Dames and Moore, Harza, HDR, INCA, and Montgomery Watson. He received his B.A. and M.A. degrees in biology from Humboldt State University and his Ph.D. degree in fisheries from the University of Washington.

HELEN M. INGRAM is a professor of social ecology and the Drew, Chace and Erin Warmington Chair in the Social Ecology of Peace and International Cooperation at the University of California, Irvine. Her research interests include transboundary national resources, particularly on the U.S.-Mexico border, water resources and equity, public policy design and implementation, and the impact of policy on democracy and public participation. Dr. Ingram received her Ph.D. degree from Columbia University.

W. CARTER JOHNSON is a professor of ecology in the Deparment of Horticulture, Forestry, Landscape, and Parks at South Dakota State University in Brookings, South Dakota. Before his post at South Dakota State University, Dr. Johnson was with the Department of Biology at Virginia Polytechnic Institute and State University. Dr. Johnson's research interests are in streamflow regulation and riparian ecosystems, restoration of ecological and economic sustainability of western rangelands, and global climate change and prairie wetlands. He received the W.S. Cooper Award in 1996 from the Ecological Society of America. Dr. Johnson served as a member of the National Research Council Committee on the Missouri River Ecosystem Science. He received received his B.S. degree in biology from Augustana College and his Ph.D. degree in botany from North Dakota State University.

JOHN J. MAGNUSON is professor emeritus of zoology and limnology at the University of Wisconsin. He is past director of the university's Center for Limnology and North Temperate Lakes Long-term Ecological Research Program. Dr. Magnuson's research interests include long-term regional ecology, climate change effects on lake ecological systems, fish and fisheries ecology, and community ecology of lakes as islands. He received his B.Sc. and M.Sc. degrees from the University of Minnesota and his Ph.D. degree in zoology with a minor in oceanography from the University of British Columbia.

STUART W. MCKENZIE is retired from the U.S. Geological Survey. While with the USGS in Maryland, he studied the saltwater interface movement in Maryland coastal streams and long-term trends of water quality in Delaware. When at the University of Delaware, he helped assess groundwater resources in the Dover area. In Oregon he has studied quality of urban runoff and impacts of agricultural runoff on streams and worked on the Willamette River Quality Assessment, Yakima River Basin Water Quality Assessment, and Intergovernmental Task Force on Monitoring Water Quality. He is currently compiling and evaluating water temperature data from across the Columbia River. He received his B.S. degree in physics from the University of Puget Sound and his MCE degree in civil engineering from the University of Delaware.

DIANE M. MCKNIGHT is a professor in the Department of Civil, Environmental, and Architectural Engineering and a fellow of the Institute of Arctic and Alpine Research at the University of Colorado. Dr. McKnight was a research scientist at the U.S. Geological Survey, Water Resources Division. She studies biogeochemical processes, aquatic ecology, and reactive solute transport in streams and lakes in the Rocky Mountains and in polar desert areas of Antarctica. Dr. McKnight is the acting president of the biogeosciences section of American Geological Union. She is currently a member of U.S. Global Climate Research Program and International Panel on Climate Change committees on climate change and water resources. Her research interests are in limnology and biogeochemical processes in natural waters. She received her B.S. degree in mechanical engineering, her M.S. degree in civil engineering, and her Ph.D. degree in environmental engineering from the Massachusetts Institute of Technology.

TAMMY J. NEWCOMB is the Lake Huron basin coordinator for the Michigan Department of Natural Resources Fisheries Division in Lansing, Michigan. In this position she coordinates ecosystem and watershed management for the Lake Huron drainages and the Lake Huron sport, tribal, and commercial fisheries. Dr. Newcomb is also an adjunct faculty member at Virginia Polytechnic Institute and State University with a research focus on salmonid population dynamics, watershed and stream habitat management, and stream temperature modeling. Dr. Newcomb received her Ph.D. degree from Michigan State University.

KENNETH K. TANJI is professor emeritus of hydrologic science in the Department of Land, Air and Water Resources at the University of California, Davis. He is a fellow of three professional societies. His major research interest is on water quality aspects of irrigation and drainage. Dr. Tanji has served on three previous National Research Council committees. He received his B.A. degree in chemistry from the University of Hawaii, Honolulu, his B.S. and M.S. degrees in soil science from the

University of California, Davis, and his Sc.D. degree in agricultural science from Kyoto University, Japan.

JOHN E. THORSON is an administrative law judge with the California Public Utilities Commission (CPUC) in San Francisco. An attorney, Thorson was special master for the Arizona General Stream Adjudication. He has served as regional counsel for the Western Governors' Conference, director of the Conference of Western Attorneys General, consultant to many state governments and courts; and director of the Missouri River Management Project for the Northern Lights Institute. Dr. Thorson received his B.A. degree from the University of New Mexico, his J.D. degree in 1973 from the University of California, Berkeley, and his doctorate in public administration from the University of Southern California.

STAFF

JEFFREY W. JACOBS is a senior program officer with the National Research Council's Water Science and Technology Board. His research interests include policy and organizational arrangements for water resources management and the use of scientific information in water resources decision making. He has studied these issues extensively in both the United States and mainland Southeast Asia. Since joining the NRC in 1997, he has served as study director for 13 study committees. He received his B.S. degree from Texas A&M University, his M.A. degree from the University of California, Riverside, and his Ph.D. degree from the University of Colorado.

DAVID POLICANSKY is the associate director of the Board on Environmental Studies and Toxicology at the National Research Council. Formerly, he taught and did research at the University of Chicago, the University of Massachusetts at Boston, and the Grey Herbarium of Harvard University. He was visiting scientist at the national Marine Fisheries Service Northeast Fisheries Center. He is a member of the Ecological Society of America, the American Fisheries Society, and the advisory councils to the University of Alaska's School of Fisheries and Ocean Sciences and the University of British Columbia's Fisheries Centre. He was a member of the editorial board of *Bioscience*. His interests include genetics, evolution, and ecology, particularly the effects of fishing on fish populations, ecological risk assessment, and natural resource management. He received his B.A. in biology from Stanford University and his M.S. and Ph.D. degrees in biology from the University of Oregon.

ELLEN A. DE GUZMAN is a research associate with the National Research Council's Water Science and Technology Board. She has worked on a number of studies including Privatization of Water Services in the United States, Review of the USGS National Water Quality Assessment Program, and Drinking Water Contaminants (Phase II). She co-edits the WSTB newsletter and annual report and manages the WSTB homepage. She received her B.A. degree from the University of the Philippines.

Appendix F

National Research Council Board Membership and Staff

WATER SCIENCE AND TECHNOLOGY BOARD

RICHARD G. LUTHY, *Chair,* Stanford University, Stanford, California
JOAN B. ROSE, *Vice Chair,* Michigan State University, East Lansing
RICHELLE M. ALLEN-KING, University at Buffalo (SUNY), Buffalo, New York
GREGORY B. BAECHER, University of Maryland, College Park
KENNETH R. BRADBURY, Wisconsin Geological and Natural History Survey, Madison
JAMES CROOK, Water Reuse Consultant, Norwell, Massachusetts
EFI FOUFOULA-GEORGIOU, University of Minnesota, Minneapolis
PETER GLEICK, Pacific Institute for Studies in Development, Environment, and Security, Oakland, California
JOHN LETEY, JR., University of California, Riverside
CHRISTINE L. MOE, Emory University, Atlanta, Georgia
ROBERT PERCIASEPE, National Audubon Society, Washington, DC
JERALD L. SCHNOOR, University of Iowa, Iowa City
LEONARD SHABMAN, Virginia Polytechnic Institute and State University, Blacksburg
R. RHODES TRUSSELL, Trussell Technologies, Inc., Pasadena, California
KARL K. TUREKIAN, Yale University, New Haven, Connecticut
HAME M. WATT, Independent Consultant, Washington, DC
JAMES L. WESCOAT, JR., University of Illinois at Urbana-Champaign

Staff

STEPHEN D. PARKER, Director
LAURA J. EHLERS, Senior Staff Officer
JEFFREY W. JACOBS, Senior Staff Officer
WILLIAM S. LOGAN, Senior Staff Officer
LAUREN E. ALEXANDER, Staff Officer
MARK C. GIBSON, Staff Officer
STEPHANIE E. JOHNSON, Staff Officer

M. JEANNE AQUILINO, Administrative Associate
ELLEN A. DE GUZMAN, Research Associate
PATRICIA JONES KERSHAW, Study/Research Associate
ANITA A. HALL, Administrative Assistant
DOROTHY K. WEIR, Senior Project Assistant

BOARD ON ENVIRONMENTAL STUDIES AND TOXICOLOGY

JONATHAN M. SAMET, *Chair*, Johns Hopkins University, Baltimore, Maryland
DAVID ALLEN, University of Texas, Austin
THOMAS BURKE, Johns Hopkins University, Baltimore, Maryland
JUDITH C. CHOW, Desert Research Institute, Reno, Nevada
COSTEL D. DENSON, University of Delaware, Newark
E. DONALD ELLIOTT, Wilkie, Farr & Galagher, LLP, Washington, DC
CHRISTOPHER B. FIELD, Carnegie Institute of Washington, Stanford, California
WILLIAM H. GLAZE, Oregon Health and Science University, Beaverton
SHERRI W. GOODMAN, Center for Naval Analyses, Alexandria, Virginia
DANIEL S. GREENBAUM, Health Effects Institute, Cambridge, Massachusetts
ROGENE HENDERSON, Lovelace Respiratory Research Institute, Albuquerque, New Mexico
CAROL HENRY, American Chemistry Council, Arlington, Virginia
ROBERT HUGGETT, Michigan State University, East Lansing
BARRY L. JOHNSON, Emory University, Atlanta, Georgia
JAMES H. JOHNSON, Howard University, Washington, DC
JUDITH L. MEYER, University of Georgia, Athens
PATRICK Y. O'BRIEN, ChevronTexaco Energy Technology Company, Richmond, California
DOROTHY E. PATTON, International Life Sciences Institute, Washington, DC
STEWARD T.A. PICKETT, Institute of Ecosystem Studies, Millbrook, New York
ARMISTEAD G. RUSSELL, Georgia Institute of Technology, Atlanta
LOUISE M. RYAN, Harvard University, Boston, Massachusetts
KIRK SMITH, University of California, Berkeley
LISA SPEER, Natural Resources Defense Council, New York
G. DAVID TILMAN, University of Minnesota, St. Paul
CHRIS G. WHIPPLE, Environ, Inc., Emeryville, Caflironia

LAUREN A. ZEISE, California Environmental Protection Agency, Oakland

Senior Staff

JAMES J. REISA, Director
DAVID J. POLICANSKY, Associate Director
RAYMOND A. WASSEL, Senior Program Director for Environmental Sciences and Engineering
KULBIR BAKSHI, Program Director for Toxicology
ROBERTA M. WEDGE, Program Director for Risk Analysis
K. JOHN HOLMES, Senior Staff Officer
SUSAN N.J. MARTEL, Senior Staff Officer
SUZANNE VAN DRUNICK, Senior Staff Officer
EILEEN N. ABT, Senior Staff Officer
ELLEN K. MANTUS, Senior Staff Officer
RUTH E. CROSSGROVE, Managing Editor